GOVERNMENT AND NORTH SEA OIL

GOVERNMENT AND NORTH SEA OIL

Danny Hann

MACMILLAN

First published 1986

Published by
THE MACMILLAN PRESS LTD
Houndmills, Basingstoke, Hampshire RG21 2XS
and London
Companies and representatives
throughout the world

Printed in Hong Kong

British Library Cataloguing in Publication Data
Hann, Danny
Government and North Sea oil.
1.Offshore oil industry — Government policy —
Great Britain — History
I.Title
338.2'7282'0916336   HD9571.6
ISBN 0–333–39436–4

To my family

Contents

Acknowledgements

I would like to thank the members of the Economics Department at the University of Surrey for providing an encouraging environment in which to undertake academic research. A particular debt of gratitude is owed to Professor Colin Robinson for his advice and guidance throughout this project. My thanks are also extended to the Shell Trust for Higher Education for their generous financial support.

In addition, thanks go to Ms S. Williams for her skill and diligence in typing this work.

DANNY HANN

1 Introduction

1.1 OBJECTIVES AND METHODOLOGY

There are two key objectives of this book. UK oil policy since the passage of the 1964 Continental Shelf Act is examined and a positive appraisal of policy is made. The UK government process is analysed, with respect to North Sea oil, by focusing on the behaviour of various groups and individuals involved in the policy process. The government is not viewed as a single entity with one set of objectives and priorities. Government is a dynamic process and policy influences and decisions are made in a constantly changing political and economic environment. An assessment of the government policy process is necessary in order to provide an understanding of the nature of the domestic oil industry and its relationship with the government.

The method by which a positive evaluation of the government oil policy process is achieved is by employing a theoretical framework provided by the economic theories of politics and bureaucracies. The works of Downs, Breton, Tullock, Buchanan and Niskanen provide the methodological structure which is adapted and applied to the development of North Sea oil policy. The notion of an omniscient and altruistic government is naïve and simplistic and is the basis of much theoretical work advocating government intervention to correct situations in an imperfect industry.[1] A recurrent theme in this study is that the government itself functions imperfectly. A basic condition for state intervention in an industry is that the industry is imperfect, for example, an oligopolistic industry where all costs are not internalised. However, a second condition must also be satisfied; that the government is able to improve on the existing situation. The rôle of the government is thus examined in the context of an imperfect government as well as an imperfect oil industry. Individuals and groups within the policy process possess utility functions which they attempt to maximise. For example, the government bureaucrat is not necessarily a political neutral and the politician is not necessarily willing to subjugate his or her own ambitions for the sake of some

1

government objective. Assumptions of bureaucratic and political behaviour (i.e. rationally attempting to maximise a personal utility function) are analogous to those made of individuals in the private sector.

Problems associated with the formulation and development of policy in the UK system of representative democracy may be seen to fall into two broad areas. First, because of the inherent shortcomings of the UK voting system the policy preferences and desires of the voting population are not clearly recognised by politicians. Thus, any 'national interest' objective (if one can exist) as determined by voters, is unlikely to be revealed to policy-makers. Secondly, even if the politician is able to identify a genuine (i.e. voter-determined) 'national interest' objective, a policy then has to be developed and implemented so as to achieve that objective. Various groups and individuals within the policy process, both inside and outside government, exert pressure on the development of policy in order to further or protect their own interests as they perceive them. Thus those active in the policy process may desire certain characteristics of policy rather than aspects of the policy outcome.

Whilst economic theory provides the framework of analysis, the political economy of North Sea oil is an integral part of the study. It is important to note that the government has numerous functions to perform in society to do with social, political and economic activities. Politicians may implement a policy for political reasons, subordinating economic considerations. This is seldom made clear and it is important that the economic implications of government are known in order to contribute to the ability of voters in formulating their policy preferences.

The intrinsic deficiencies of the government policy-process as examined in the context of domestic oil policy thus have an important bearing on the government's ability to improve on the functioning of the offshore oil industry.

1.2 STRUCTURE

The historical background to the development of UK North Sea oil policy is outlined in Chapter 2. The discovery of oil on the United Kingdom Continental Shelf (UKCS) in the late 1960s and 1970s had the effect of changing the status of Britain from an oil importing nation to a net oil exporter in the 1980s. In 1982 the UK was producing over two million barrels of oil per day, roughly equivalent to middle-ranking OPEC producers.[2] The historical development of the UK oil sector since 1964 provides the background to policy decisions made in the 1970s and 1980s. The timing of various national and international events (both political and economic) have had an important effect on the UK oil sector in two major ways. First, events such as the world oil price increases of 1973–4 and of 1979 had a direct effect on the economics of North Sea oil. Second, these events additionally have an indirect effect on the UK oil sector because they alter perceptions of the oil market and the importance of UK oil as a political issue. Political events in the UK can similarly be seen to affect oil policy. For instance, the changing political complexion of the government, macro-economic priorities of governments and the timing of General Elections, all have an effect on the relationship between government and the oil industry.

Chapter 3 provides the theoretical framework for the analysis of government and North Sea oil policy. Much of the work in this chapter is derived from work presented in the context of the US system of government. The various theories of politics and bureaucracies are thus selected for, or adapted to, application for the UK system of government. Over time, according to various political and economic factors, different government departments (for example, the Foreign Office, Department of Energy, the Treasury) have varying degrees of influence on the development of oil policy. The impacts of pressure groups, political rivalries and voting rules on the policy process are outlined in this chapter.

The economic methodology set out in Chapter 3 is applied to North Sea oil policy in the subsequent four chapters. A detailed account is given of the evolution of licensing, taxation, participation and depletion policies. The economic theories of politics and bureaucracies are employed in an attempt to make a positive assessment of oil policies in the UK and to highlight various distortions and inconsistencies in the oil policy process.

Chapter 8, the final synthesising chapter, provides an overview of UK oil policy. The implications of the analysis are outlined by comparing the traditional rôle of the government in the economy to the experience of government involvement in the North Sea oil industry. Suggestions concerning the policy process are referred to and future policy trends are considered in the context of the economic theories of politics and bureaucracies.

2   The Historical Background

2.1 INTRODUCTION

The development of North Sea oil and gas over the last two decades has, in many respects, been highly successful. Oil companies have overcome many technical problems of deep sea drilling which had never before been confronted. The first offshore drilling took place in Dutch coastal waters in 1961 and following the passing of the 1964 Continental Shelf Act, the first gas in the British sector was discovered in the West Sole Field in 1965. This was followed by further discoveries in 1966 of gas in the Leman Bank, Indefatigable and Hewett Fields. These and further significant discoveries (enough to support a 'plateau' output of around 4000 million cubic feet per day throughout the 1970s and 1980s)[1] led to the conversion of domestic consumers to natural gas. Between 1970 and 1978 UK consumption of natural gas increased from 4400 million therms to 16 500 million therms[2] each year. As exploration drilling moved further northward, towards the end of the 1960s important oil reserves were discovered. In 1971 BP's Forties Field was declared commercial, as were Auk, Brent and Argyll in 1972; Argyll and Forties commenced production in 1975. At a time of rapidly increasing world oil prices in the first half of the 1970s discovery rates peaked and by 1980 total (proven, probable and possible) recoverable reserves on the UKCS were estimated to be between 2200 and 4400 million tonnes.[3] Thus, during the 1980s the UK would become a net oil exporter producing as much oil as Venezuela, Kuwait and Nigeria.

This chapter attempts to provide a background to the overall development of UK offshore oil and rôle the government has adopted since 1964. The influence of the government on the offshore oil industry has been considerable and often controversial. The policies of successive governments must be seen as an integral part of the problem of development of North Sea oil. Not unlike the experience of other oil producing regions, government policy has changed considerably from a relative absence of intervention and a fiscal régime encouraging exploration and development when reserves were unproven and risks were high, to a policy of increasingly stringent controls and increasing government intervention as substantial reserves were declared commercial. In outlining the historical development of the relationship between the government and the offshore oil industry three distinct periods can be observed.[4] Each period may be characterised by discernible policy trends which have affected the economics of domestic offshore oil. In briefly highlighting overall policy trends concerning North Sea oil this chapter provides the background to subsequent chapters examining North Sea oil policy in greater detail.

2.2 1964–1974

The 1964 Continental Shelf Act served to extend the oil and gas licensing and regulatory powers granted to the government by the 1934 Petroleum Production Act to offshore areas. The government's basic policy considerations at this time were relatively uncomplicated. A climate was to be provided which would encourage the most rapid and thorough exploration and development of the UKCS at a time when the oil companies were naturally cautious because of the enormous risks involved and because massive reserves were cheaply available elsewhere, notably in the Middle East. The Conservative Government was concerned about inciting OPEC, at a time when OPEC was pressing for increased revenue from oil company concessionaires, to worsen the terms and harm UK overseas oil interests and the UK Balance of Payments in response to excessively burdensome terms imposed by the UK Government on oil companies operating in the North Sea. The government was also considering the sizeable benefits from the offshore supplies industry which could accrue to UK firms whether or not reserves of oil or gas were found. A recurring factor throughout this period which had a significant effect on the organisational framework in which North Sea activities evolved was the government's desire for speed. This desire to promote a rapid rate of exploration (due largely to the poor performance of the post-war domestic economy)[5] has been cited[6] as the primary reason why the UK Government did not contest the division of the North Sea between the UK and Norway which resulted in the UK obtaining only 35 per cent of the North Sea.

Although the government's rôle in the North Sea at this time was characterised by a relative lack of intervention, it did make decisions as to the structure of control which had far-reaching implications. The decision was made to employ a discretionary licensing system. In doing so, the government set the framework in which future government intervention and government-bureaucrat control could occur. The discretionary licensing system and its implications are examined in Chapter 4 but it is important to note that during this period of relative absence of government intervention, the government ensured that the potential for state control existed in the early legislative framework.

The First Licensing Round in 1964 allocated 53 licences covering 348 blocks.[7] The General Election of October 1964 brought the Labour Party into power and following indications of gas in the Southern Basin, the new government was keen to give exploration an added impetus by announcing in 1965 the Second Licensing Round. Soon after the announcement of the Second Licensing Round, BP discovered gas in what became known as the West Sole Field: this was followed in 1966 by further discoveries in the Southern Basin. Experience in other areas suggested that it would be unusual if no more gas were found and thus in just over two years the southern part of the North Sea had changed from a completely unproven area to one of considerable significance as a gas producing basin. This prompted the Government to institute a major review of future licensing procedure and of general policy with respect to this new domestic source of energy. The Labour Party Fuel Study Group in 1968 proposed the setting up of a National Hydrocarbons Corporation to assume sole responsibility for exploration and development in all the offshore areas not retained by existing licenses. This proposal was not followed through but it did clearly illustrate the Labour Party's commitment to some active state corporation in the North Sea foreshadowing the British National Oil Corporation and indicating the government's undertaking to increase state participation in the future. Other forms of participation, as in the Netherlands, Norway or Denmark were rejected because of the need to develop smaller, higher cost gas fields and the continuously stressed need for speed.

Following the discovery of the small uncommercial Cod Field in 1969, the huge Ekofisk oil field was discovered in the Norwegian sector. This, as well as the low gas price, prompted the shift further north and an intensification of the search for oil, and in 1970 BP discovered the Forties Field in the British sector. At the same time there was increasing pressure being applied by OPEC members on concessionary companies. In September 1970 Libya was successful in increasing its take from oil companies and other countries began to follow the Libyan lead. An estimated additional £200m[8] in exchange costs to the UK from oil imports had been incurred and there was now greater attention paid to promoting secure oil supplies. Together with the uncertainties of the Middle East and early discoveries of oil in the North Sea there was a third factor which influenced the government's decision to announce a Fourth Licensing Round. Due largely to the small size of the Third Round there had been a significant slowdown of activity in the North Sea.

Thus at this time the government was beginning seriously to consider the attractions of a new, more involved rôle for itself in the North Sea. However, its two main policy considerations were still to secure as quickly as possible the maximum effort of exploration and development of the UKCS and to ensure British interests were adequately protected. The Fourth Round of licensing was successful in speeding up the development of the UKCS (as were increasingly fast moving events in the Middle East) and there was some concern that the offshore supplies industry could not cope with the increased demand of the companies operating in the much fiercer environment of the northern North Sea. There had been some criticism of the low share (25–30 per cent in 1972) of the British offshore supplies industry in the North Sea and it was significant that during this period of high activity British shipyards failed to take advantage of the surge in rig orders. Offshore engineering and contracting were mainly in the hands of American companies who tended to favour familiar and proven equipment from known American suppliers.

To the extent that development in the North Sea had occurred relatively quickly, the government's rapid exploitation policy had been successful (see Tables 2.1 and 2.3). Subsequent chapters will examine the wider implications of this policy, but by the early 1970s economic and political conditions had changed. The government was keen to increase its take and also its powers of control. Due to the uncertainties involved and the need to attract oil companies into the North Sea, in 1964 the government was in a relatively weak bargaining position and was prepared to refrain from intervention, enforced no explicit production controls and maintained a relatively lenient tax system. The only significant area of government involvement in the

TABLE 2.1　*UK offshore drilling activity*

| | Number of wells drilled each year | | |
	Exploration	*Appraisal*	*Development*
1964	1	0	0
1965	10	0	0
1966	20	8	3
1967	42	16	13
1968	31	8	36
1969	44	8	27
1970	22	2	28
1971	24	4	34
1972	33	8	36
1973	42	19	21
1974	67	33	20
1975	78	37	21
1976	58	28	54
1977	67	38	96
1978	37	25	96
1979	33	15	102
1980	32	22	122
1981	48	26	137
1982	68	43	118
1983	77	51	96
1984	106	76	108

SOURCE　Department of Energy, *Development of the Oil and Gas Resources of the United Kingdom 1975 to 1985.*

North Sea was with respect to gas. Largely as a consequence of the discretionary licensing system (see Chapter 4), the government realised that if gas were to be sold by the oil companies at market prices, economic rent would be passed to the companies in the form of profit. In order to capture the economic rent the Gas Council was made the sole buyer of offshore natural gas, companies were prohibited from selling direct to industry and from direct exports.[9] The Gas Council (and its successor, the British Gas Corporation) thus acquired monopsony and monopoly powers. The gas·price negotiated with the companies was established at a very low 1.2 pence per therm but it is unlikely that the economic rent was passed on to the consumer or captured by the government by taxing BCG's profits.[10]

Due to the absence of competition, it could be predicted[11] that considerable 'X-inefficiency' would develop within the BGC.

2.3 1974–1979

In the 1960s the real price of oil was falling. Thus $\dot{p}$, defined as the expected rate of appreciation of the net price of oil, was negative. Also, because multi-national oil companies predicted that at some time in the future they could lose control of some of their oil activities, most notably in the Middle East, oil company discount rates (i) tended to be relatively high. An oil producer will attempt to equate the opportunity cost of production ($\dot{p}$) to the opportunity cost of not producing i in order to maximise Net Present Value at the rate of discount. Thus in 1960s, $\dot{p} < i$ and output in the short term increased. In the 1970s as oil producing countries took control of production rates and prices were expected to rise, the oil producers had longer time horizons than the companies and the relationship reversed; $\dot{p} > i$. There was an incentive for oil producers to hold back production in the short term as, given their discount rate, oil producers could expect to maximise NPV by investing in oil in the ground rather than extracting it in the short term.[12] At the same time there was an influential view that as consumption of oil increased, fuel shortages would develop.[13] Although this argument seems to ignore the effects of oil price increases on oil substitutes and the interaction of demand and supply as prices change, there was considerable support for some sort of 'energy-gap' theory.

Nevertheless, the quadrupling of the world price of oil in the winter of 1973–4 had a significant effect on the UK Government's perception of North Sea oil. Apart from the balance of payments effects, the most serious problem for Britain was perceived to be the potential loss of secure supplies of oil. Even with the Government's shareholding in BP it could not persuade BP to favour Britain in its allocation of oil. There were significant policy implications for North Sea oil. The licensing terms did not give the government explicit control over either the rate of extraction or the destination of the oil once it had been landed in the UK. Thus by 1974 a combination of events resulted in the reversal of the rapid exploitation policy. Bureaucratic pressures and political pressures from both within the UK and internationally necessitated fundamental changes in government offshore oil policy. These multifarious pressures will be explicitly examined in later chapters.

Following the fall of the Heath Government the new Labour Administration instituted a review of oil policy which produced an influential White Paper.[14] Because of higher oil prices the existing tax and royalty arrangements were not considered adequate and it was estimated that unless these conditions were changed the government would never take much more than one half of the oil company profits, and about one half of the post-tax profits would be remitted overseas.[15] Thus the government was keen to increase its take and because of higher oil prices the government's priority of rapid exploitation was now being superseded by other considerations.

In the White Paper the government outlined its two principal objectives. Firstly, 'to secure a fairer share of profits for the nation and to maximise the gain to the balance of payments', and secondly, 'to exert greater public control' so as to 'safeguard the national interest'.[16] In order to achieve these objectives the White Paper proposed five measures, three of which related directly to the state participation in the exploitation of oil resources on the UKCS. It would be a condition of future licences that, if required to do so by the government, the licensees should grant majority participation to the state in all fields discovered under these licences. The Labour Government believed that majority state participation in existing licences provided the best means for the nation to share fully in the benefits of North Sea oil without 'unfairness' to the licensees as the state contributed its share of the costs, including past costs. A British National Oil Corporation (BNOC) would be set up through which the Government would exercise its participation rights. BNOC would represent the government in the present consortia, and would take over the National Coal Board's interests. BNOC would play an active part in the future exploration, development and exploitation of the UKCS and would have powers to expand its activities downstream and abroad. The other measures the government proposed in the 1974 White Paper were for a Finance Bill which would impose an additional tax on the companies' profits from the UKCS. Also the government would extend its powers to control physical production and pipelines: to 'take power to control the level of production in the national interest'. Furthermore, the government would require licensees to provide more information about their activities than was at that time obligatory.[17]

Thus the 1974 White Paper clearly stated the Labour Government's general policy intentions with respect to the North Sea. In parliamentary debates following the publication of the 1974 White Paper, Eric Varley (the Secretary of State of Energy) and John Smith (Under-Secretary) justified the government's participation policies and the establishment of BNOC.

Although the 1964 Continental Shelf Act had already established state ownership of North Sea reserves, it was still Labour Party policy to honour its commitment stated in its 1974 election manifesto. This was 'to ensure not only that the North Sea and Celtic Sea oil and gas resources are in full public ownership, but that the operation of getting and distributing them is under full government control with majority participation'.[18] Thus BNOC was to be the vehicle by which the socialist principle of national ownership would be implemented. The 1974 Labour Government had therefore emphasised the importance of North Sea oil as a political issue. It had become a major part of the Government's proposed parliamentary programme and election platform. North Sea oil policy was now subject to all the pressures and uncertainties that attach to an important element of government policy, to a far greater degree than it had been in the 1960s.

In the debate on the second reading of the Petroleum and Submarine Pipelines Bill (which was officially to establish BNOC) in April 1975, it was significant that Labour members frequently referred to the Bill as being a truly socialist bill, 'if one accepts, as Labour Members do, that it (state intervention in any industry) is a proper function of the nation, there is surely no better use than that of oil'.[19] The Petroleum and Submarine Pipelines Act, in setting up BNOC, must be seen primarily as an Act which attempted to further the state's influence in a major area of the economy. Critics have tended to view BNOC's proposed 51 per cent participation as the first step towards wholesale nationalisation. It is probably more accurate to say that 51 per cent participation was included in BNOC's brief in order to satisfy the left wing of the Labour Party and was never intended to be rigidly adhered to. This is an example of intra-party rivalry affecting the policy process and will be examined in greater detail in subsequent chapters.

A second and less doctrinaire objective of government policy with respect to BNOC and North Sea oil was to 'acquire our own direct knowledge of the difficult techniques of oil and gas production'.[20] Both major parties agreed on the necessity to enhance the government's access to information and its ability to interpret that information. The two parties differed in that the Opposition felt that this could most efficiently be brought about not by the setting up of a vast and expensive state corporation but by establishing some sort of regulatory agency. The government's view was that access to information was only one of the functions BNOC would fulfil.[21] The

Conservative proposal was for a UK Oil Conservation Authority, (UKOCA) which would be 'a small body, of no more than five members, though with power to take on technical assessors for particular studies'.[22] UKOCA would be based on the pattern of the Energy Resources Conservation Board of Alberta, thus acting only as an impartial watchdog. It was not considered by the government as an alternative to BNOC. Tony Benn (the successor to Eric Varley) justified the need for active and large-scale participation on the grounds that first-hand experience is vital to gain the necessary expertise and information in all North Sea activities. The much quoted argument is that North Sea oil is too important to the economy as a whole, first to be left to free market forces, and secondly to base crucial policy pronouncements on information de- rived indirectly: 'one cannot live off decisions of that magnitude and be dependent on the advice of others'.[23]

A third government objective closely associated with the need for information was the need for government control over North Sea oil development and disposal. These were principles accepted by both parties but the method by which they were to be achieved differed between the parties. Extensive powers were given to the Secretary of State for Energy to regulate production in order to promote and protect the government's notion of 'national interest'. The govern- ment intended that once self-sufficiency had been reached in the early 1980s, strict production controls would be enforced and the Petro- leum and Submarine Pipelines Act would provide the means by which future controls could be implemented. Thus policies of rapid exploitation now seemed to be completely in the past, especially as the Conservative Party agreed with the need to have government depletion controls, also seeming to disregard the effect of market signals on oil companies' production profiles. In response to oil company complaints of increasing uncertainty, Eric Varley an- nounced guidelines covering the implementation of the Minister's powers on production controls. Thus the so-called 'Varley Guide- lines' outlined by the Energy Secretary were an attempt to assure the oil companies that the extensive powers on depletion rates given to the Minister in the Petroleum and Submarine Pipelines Act would not 'undermine the basis on which they have made plans and entered into commitments'.[24] However, the guidelines did not achieve this, largely because uncertainties remained and the guidelines could be interpreted in various ways. There was doubt about the base figure to which the percentage production cut limits applied and confusion as to whether governments would discriminate between fields or hold back new developments.[25]

Mr Varley's guidelines were as follows:

(1) On finds made up to the end of 1975 there would be no delays imposed on the development plans. If delays were to be imposed on the development of finds in or after 1976 the government would act to prevent companies from making premature investments.

(2) On finds made up to the end of 1975 there would be no cuts in production before 1982 or until four years after the start of production, whichever was later.

(3) On finds made after 1975 on an existing licence no cuts would be made in production until 150 per cent of the capital investment in the field had been recovered.

(4) When using their depletion powers the government would consider technical and commercial factors associated with individual fields which would generally limit cuts to 20 per cent. The industry would be consulted on the period of notice to be given before any cuts would come into effect.

(5) In deciding on production cuts the government would take account of the needs of the UK offshore supplies industry.

Further government controls on oil company activities, as stated in licence agreements, included the provision that all North Sea oil and gas should be landed in the UK even if it was to be subsequently exported.[26] This would have the effect of increasing transport costs for a company whose field is not on a pipeline to the UK mainland and therefore could have exported its oil directly by tanker. The government also ruled that two-thirds of North Sea oil should be refined in the UK. This rule was not strictly adhered to both because of the potential Balance of Payments loss, and because it would have depressed the price of oil domestically as refinery capacity would not be able to cope. Although these controls were left open to considerable discussion between the government and oil companies they were symptomatic of the government's desire to increase its sphere of influence in all aspects of the North Sea oil industry.

A further objective of state participation as seen by the Labour Government was that it was an effective way of increasing oil revenues. The 1975 Oil Taxation Act's effect was to split profits 70:30 in the government's favour and with BNOC's 51 per cent participation this ratio would shift to 85:15 after the Fifth Licensing Round.

Petroleum Revenue Tax (PRT) was introduced largely in response to a popular perception of 'windfall' profits accruing to oil companies as a result of the 1973–4 world oil price increases. PRT was set at a flat rate of 45 per cent of income net of royalties and operating expenses. To guarantee the companies a minimum return on investment capital PRT was not imposed until the original capital expenditure plus an Uplift of 75 per cent was earned. This Uplift was designed to compensate for interest payments on loans which were not deductable. In 1978 the government proposed an increase in PRT to 60 per cent, the oil production allowance (of 1 million tonnes, set in 1975) was reduced to 1/2 million tonnes per year and the Uplift was to be reduced to 25 per cent. The 1979 Conservative Government actually implemented most of the proposed tax changes; however, the Uplift was only reduced to 135 per cent (see Chapter 7). The chapter on Licensing Policy (Chapter 4), analyses the relationship between the discretionary licensing system and the tax system through which the government hoped to capture the economic rent that had been transferred to the oil companies due to the discretionary system.

Thus with the establishment of BNOC the Government's rôle in the North Sea altered considerably. Confusion and uncertainty increased as the precise rôle of BNOC was unclear as was the specific relationship between the government and BNOC. It was important to the government to have the ability to maintain flexibility in order to be able to adapt and adjust to an ever changing economic and political environment. Politicians of both major parties favoured the development of a fully integrated energy policy (national and/or international) within which oil policy would play a central part. The Petroleum and Submarine Pipelines Act can be seen as the Act which laid the legal and institutional foundations upon which the government's oil policy could develop and which had adequate scope within which the government could respond to unforeseen circumstances.[27]

One of the consequences of this institutional rather than specific forward planning was that BNOC did not have clear-cut long-term objectives. In its First Report[28] BNOC outlined its wider aims and viewed its position as combining the functions of an instrument of national policy, of a commercial enterprise and of an advisor to the government.

The original intentions with regards to participation arrangements had been for a voluntary 51 per cent participation, with BNOC paying for past and future expenditure on a 51 per cent basis. This proved totally unacceptable to the oil companies who saw 51 per cent as complete control. Following lengthy discussions with the companies a compromise was reached based around the Department of Energy's desire for information and knowledge and their desire to have a say in how the companies disposed of the oil they produced. Thus participation was re-defined to mean that BNOC was to be treated as a partner in that BNOC would have access to information but would have no equity share. 51 per cent oil was conceded to BNOC on the understanding that BNOC would hand back to the companies all financial and other benefits (i.e. the oil would be sold to BNOC at the market price). BNOC would be a member of the operating committees that manage fields but 'there was no question of exercising a 51 per cent vote'.[29] It was from this general basis that Lord Kearton conducted participation negotiations with individual companies.

The establishment of BNOC on 1 January 1976 came at a time when production of oil on the UKCS was rapidly increasing (see Table 2.2). Between 1976 and 1979 BNOC also expanded its activities; in its first official year of existence BNOC's capital expenditure amounted to £396m out of which £287m was used to acquire the North Sea interests of the National Coal Board and Burmah Oil. By the end of 1976 BNOC was a licensee in 22 production licences covering 59 blocks or part blocks, extending to more than 9000 square kilometres of the UKCS. BNOC also had equity interests in one producing gas field (Viking), five oil fields under development (Thistle, Ninian, Dunlin, Statfjord and Murchison) and associated pipeline and terminal facilities. BNOC was also operator for the Thistle Field development and for nine exploration blocks. By the end of 1976 the Corporation employed 388 people.[30]

With Government support BNOC expanded its North Sea activities considerably during the period 1976–9. In 1978 BNOC had started trading in equity, participation and third party crude oil, reaching an average level of 170 000 b/d in sales to third parties by the end of the year. BNOC was an equity partner with over 80 companies including each of the world's seven majors. In addition to its equity interests BNOC had secured participation agreements with 62 companies, had an equity participation for exploration work in over 50 licences, and was involved in 21 oil fields in production or under development in the North Sea.[31] BNOC showed a gross profit of £30m and an operating profit for the year of £11m after taking account of site restoration costs, depreciation of general exploration

TABLE 2.2 UK North Sea oil and natural gas liquids production *(thousand daily barrels)*

Project	1975	76	77	78	79	80	81	82	83	84*	85*
Established commercial fields											
Alwyn Nth											
Argyll	10	23	16	14	16	16	10	22	17	15	12
Auk	0	25	47	27	16	12	12	13	9	6	4
Balmoral									0	0	0
Beatrice						0	6	33	30	50	51
Beryl	0	8	62	53	97	110	96	92	78	105	160
Brae							0	0	23	110	110
Brae Nth									0	0	0
Brent	0	2	27	78	181	140	231	323	400	410	410
Buchan	0	0	0	0	0	0	18	28	31	25	45
Claymore	0	0	6	62	82	88	90	100	100	83	72
Clyde								0	0	0	0
Cormorant	0	0	0	0	1	9	4	6	17	17	11
Cormorant Nth	0	0	0	0	1	13	14	32	60	118	170
Duncan									0	1	8

Dunlin	0	0	0	14	122	110	98	76	74	60	60
Forties	12	177	413	503	503	512	468	455	448	373	306
Fulmar	0	0	0	0	0	0	0	50	118	135	135
Heather	0	0	0	2	16	14	25	33	27	19	16
Highlander											
Hutton	0		0	0	0	0	0	0	0	0	10
Hutton NW	0	0	0	0	0	0	0	0	41	55	55
Magnus				0	0	0	0	0	30	95	125
Maureen	0	2							16	70	75
Montrose	0	0	16	25	27	25	23	18	14	15	14
Murchison UK	0	2	0	1	0	8	64	92	95	99	83
Ninian	0	0	0		158	233	292	307	280	212	187
Piper	0	2	177	251	271	216	201	201	199	120	104
Statfjord UK	0	0	0	0	1	9	26	37	43	50	55
Tartan	0	0	0	0	0	0	13	13	20	22	22
Thistle	0	0	0	53	80	109	114	129	105	96	82
Totals: million barrels per day	.0	.2	.8	1.1	1.6	1.6	1.8	2.1	2.3	2.4	2.4

*Author's estimates.
SOURCE   Surrey Energy Economics Centre, University of Surrey.

costs and selling and administrative expenses. The profit attributable
to the year's operations, before taxation, was £2.3m and the net loss
for the year after writing off brought forward interest and financing
costs, and charging deferred taxation, was £3m. The majority of this
£3m was derived from fields already under development in 1976 and
1977. During 1978 BNOC became the largest world trader in North
Sea oil. The establishment and development of BNOC (see Chapter
5) was a major policy objective of the Labour Government and was
central to the government's overall policy objectives concerning
North Sea oil.

2.4 1979 TO THE PRESENT DAY

The new Conservative Government of 1979 was expected to bring
about a significant change in the rôle of the government in the North
Sea. Conservative Party criticism of BNOC had intensified throughout 1978, based mainly on the assertion that the activities of BNOC
were accelerating the departure of US companies and thus causing
serious delays in oil production and a sizeable loss in government
revenue. In addition, BNOC was criticised for undermining the
confidence of other companies which regard the 'majors' as natural
industry leaders. Labour Government and BNOC arguments in
response to these criticisms were that the large multi-national oil
companies would be leaving the North Sea irrespective of BNOC's
actions as remaining oil field projects became less financially attractive compared to opportunities elsewhere in the world. Thus the high
cost marginal fields would be left for the smaller, less capable companies and BNOC would have to have acquired the experience and skill
to fill the gap left by the departure of the majors.

Whilst the establishment of BNOC was central to the 1974–9
Labour Government's oil policy, its privatisation was central to the
oil policy of the 1979 Conservative Government. Privatisation of
BNOC was to be a political statement, designed to indicate clearly
the Government's commitment to reducing the influence of the
government in the economy. This commitment to the dismantling of
BNOC led to a great deal of speculation and uncertainty over the
new Administration's oil policy.

The government announced its intentions with respect to BNOC
including plans to sell off some of its assets; to divide it into an
exploration and development, and a trading company; and plans to

widen the ownership of BNOC, possibly by means of a North Sea oil stock.[32] These changes did not immediately occur. The economic theories of politics and bureaucracies (Chapter 3) highlight the difficulties in large-scale policy changes and the opposition within the groups involved to change which may diminish their power. Moreover, the structure and activities of BNOC (Chapter 5) had been such that there were numerous technical obstacles to its break-up. There are various reasons for the delay in privatising BNOC. BNOC was useful to the Government as a policy instrument in the North Sea. It had, as a conscious strategy, protected itself from Conservative Party ideology by working with the Government as an expert and willing 'ally'. 1979–80 was a time of political upheaval in the Middle East and of considerable oil price increases. A government controlled public corporation in the oil sector protecting the 'national interest' was politically important to the Government as a vote-maximising policy. In addition, the revenue from North Sea oil that accrued to the Treasury from BNOC proved to be invaluable to the government and its wider economic objectives. The delay in BNOC's privatisation may also been seen as a personal failure on behalf of the Secretary of State for Energy, Mr David Howell. This would partly explain his replacement by Mr Nigel Lawson who, in 1982, three years after the Conservative Party's election victory, did create a new company, Britoil, leaving BNOC with only its function as an oil trading company. The creation of the privately owned Britoil and the rationale behind the government activities concerning BNOC are examined within the framework of the economic theories of politics and bureaucracies in Chapter 5.

In 1979 expectations of government oil policy were simplistically formulated largely by the present Government's consistent non-interventionist pronouncements. With respect to BNOC and with other aspects of oil policy, changes were slow to occur. That BNOC remained as a public company even as an oil trader was significant. The Government was aware of the political importance of some form of state-controlled company in the oil sector. BNOC's usefulness to the government was well illustrated after OPEC's London Meeting in March 1983 when the UK Government, via BNOC, crucially priced North Sea oil at a level which would not encourage Nigerian retaliation.

Further expected policy changes were similarly slow to come about. The oil taxation system became more onerous over time. Changes included the increase of PRT to 70 per cent and then to 75 per cent, a supplementary oil tax (SPD) was introduced and later

abolished and Advance Petroleum Revenue Tax (APRT) was introduced (later to be phased out). These measures illustrate the government's desire for short-term revenue and its willingness to
compromise ideological and economic objectives for wider political
considerations. In 1982 minor[33] oil tax concessions were introduced
and in 1983 the Government made significant tax changes, chiefly
affecting new development projects (see Chapter 7). The 1983 tax
changes were announced in the context of intense industry pressure
for change and the slowdown of activity on the UKCS. Falling oil
prices in early 1983 also contributed to the Government's action to
promote exploration and development activity in the North Sea by oil
tax changes.

Prior to July 1980 depletion policy was concerned with maintaining
Government powers and overseeing company production plans.[34] In
the summer of 1980, Mr Howell announced government plans to
delay some development projects and regulate production in 'the
national interest'.[35] BNOC's Clyde Field project was reported[36] to
have been delayed as a result of government depletion policy. Short-
term priorities of the government are a feature which has influenced
the development and implementation of offshore oil policy over time.
BNOC, as a public corporation, was persuaded to delay the Clyde
development project for two years largely because capital expenditure would have increased the PSBR. Clyde was eventually given
development consent in 1982 when capital costs would be incurred by
the privatised Britoil. As with other aspects of offshore oil policy
(notably with respect to BNOC) the Government subsequently revised its plans and in June 1982 Mr Lawson announced[37] that there
would be no production cutbacks imposed on oil companies before 1985.

The Oil and Gas (Enterprise) Act 1982[38] which created the private
sector company, Britoil, also reduced the monopsony/monopoly rôle
of the BGC. The Act removed the obligation of companies to sell
natural gas to BGC and allowed consumers to purchase gas from any
supplier. In some circumstances (for users of 2 million therms a year
or less) the consent of the Secretary of State is required for gas to be
supplied by a company other than BGC. For users of more than 2
million therms a year the Secretary of State's consent is not required
and for users of less than 25 000 therms a year and who are within 25
yards of a BGC main, gas may be supplied by a private company if
BGC does not object to the consent given by the Secretary of State.[39]
A second change to BGC brought about by the Act was the oppor-

tunity for private companies to negotiate with BGC for the use of BGC's onshore pipeline system for the transmission of gas.

Furthermore, as part of the Conservative Government's policy of privatisation (and of raising short-term revenue), the Energy Secretary, Mr Peter Walker, instructed BGC to sell its 50 per cent share in the Wytch Farm oilfield.[40] In addition, BGC was ordered to transfer its holdings in five offshore oilfields (Beryl, Hutton, NW Hutton, Montrose and Fulmar) and its interests in eight offshore licences. The measures outlined above with respect to BGC were of considerable political value to the Conservative Government in an election period. The Government could be seen to be encouraging competition and to be reducing the powers and activities of a nationalised industry. In addition, Treasury objectives for short term revenues were achieved.

The Eighth and Ninth Licensing Rounds included auction 'experiments' but it seems that the pressures within government strongly in favour of the administratively extensive discretionary licensing system will result in the continuation of the discretionary method. The sizes of licensing rounds instituted by the Conservative Government have been irregular but their timing has been, since the Sixth Round, at more regular intervals. Special treatment for BNOC has ceased but the nationalistic consideration to give British firms preferential treatment in licence allocation has remained.[41] Licensing policy is fundamental to other aspects of offshore oil policy, notably with respect to oil taxation. Chapter 4 examines licensing policy in greater detail. Government consensus since 1964 concerning the discretionary allocation system is explained largely in terms of the economic theories of bureaucracies.

Between 1978 and 1982 oil production from the UK sector of the North Sea almost doubled. In 1983 UK oil production is expected to be approximately 2.3 million barrels per day, reaching a peak of around 2.5 m.b.d. between 1984 and 1986. Government tax revenues from North Sea oil are expected to be just under £10 billion in 1983 rising to over £14 billion in 1987 (see Table 2.3). Because of the volatile nature of the world oil market over the last decade, the offshore industry is of considerable political importance to the Government as well as for its tax revenues. Various pressure groups and bureaucracies in government and in the oil industry also have a considerable interest in oil policy and have a rôle in the development and implementation of that policy. Political and bureaucratic pressures and constraints may be seen as a common feature throughout

TABLE 2.3   Estimates of government offshore tax revenue (1975–90)

	Oil production (mm.b.p.d.)	Total revenue	Government take	PRT/APRT	SPD	CT	Royalty
1975	0.02	0.1	0.0	0.0	0.0	0.0	0.0
1976	0.2	0.9	0.04	0.0	0.0	0.0	0.02
1977	0.8	2.3	0.2	0.0	0.0	0.1	0.1
1978	1.1	2.8	0.7	0.2	0.0	0.2	0.2
1979	1.6	5.2	1.7	0.8	0.0	0.4	0.5
1980	1.6	8.8	3.0	1.6	0.0	0.6	0.8
1981	1.8	11.7	6.8	2.7	1.7	1.1	1.2
1982	2.1	14.3	7.7	3.1	2.3	0.8	1.5
1983	2.3	16.6	9.4	6.1	0.2	1.3	1.8
1984	2.4	19.2	11.2	7.0	0.0	2.1	2.1
1985	2.4	18.9	11.5	6.8	0.0	2.6	2.1
1986	2.3	19.5	12.4	7.4	0.0	2.8	2.2
1987	2.1	18.6	11.9	7.5	0.0	2.4	2.1
1988	1.9	17.8	10.4	6.6	0.0	1.9	1.9
1989	1.9	18.3	10.0	6.4	0.0	2.0	1.7
1990	1.9	19.0	9.7	6.1	0.0	2.1	1.5

Unless stated, all figures are in '000 million pounds sterling, nominal terms. Main assumptions: (i) The exchange rate is £1 = $1.35 in 1984 and £1 = $1.45 thereafter; (ii) Inflation is 5% in 1984 and onwards; (iii) the price of a 'typical' barrel of North Sea oil is assumed to be $30 in 1984 and constant in real terms thereafter; (iv) the tax system remains as of 1984.

the evolution of oil policy in the UK since 1964. These constraints and pressures have differed in character and relative strength over time as circumstances have changed. The non-interventionist policies of the 1979 Conservative Government have not occurred to the degree that was expected in 1978–9. In the context of the economic theories of politics and bureaucracies, failure to implement radical non-interventionist policies might have been predicted. The boost to exploration and development activity resulting from the 1983 Budget tax changes has been an attempt to maintain offshore activity; and uncertainty concerning UK Government oil policy since the early 1970s has not seemed to have diminished in the 1980s.

3 The Theoretical Framework

3.1 INTRODUCTION

Since the publication in 1957 of Anthony Downs' *An Economic Theory of Democracy*[1] the theory of public choice has expanded into a broader and more far-reaching analysis of the economics of politics and government decision-making. Work by economists such as Bergson, Arrow and Sen stimulated interest in the area of normative public choice and normative theorists tended to concentrate on the government's objectives being ultimately to maximise some social welfare function. Although Arrow[2] proved that given certain conditions a social welfare function will be impossible to locate there has nevertheless been a tendency to overlook the motivation of individuals involved in public decision-making and to assume conveniently that their sole purpose is to maximise some given social welfare function.

Whilst in discussion of the workings of private markets the self-interest axiom is of central importance as the basis of all hypotheses, when discussing government activities there has tended to be an implicit assumption of governments being staffed by altruists whose only concern is serving the 'public interest'. By employing economic tools (e.g. monopoly, competition, transaction costs) and applying them to politics and political institutions Downs, and subsequently other economists, attempts to construct a framework within which the decision-making process in a democracy can be studied. Thus over the last twenty-five years there has been an increasing volume of literature concerned with the intermediate stages in the decision-making process, between the initial stage of individuals making their preferences known and the final stage of a policy being implemented. It is during these intermediate stages that discrepancies and imperfections occur which can make the output of governments different from the preferences of voters. The economics of politics therefore adds to

24

the understanding of government policy in that it makes possible the examination of these intermediate stages in the decision-making process.

Within this recent literature Breton[3] classifies the writings under four broad headings: the theory of decision-rules, the theory of democracy, the theory of public goods and the theory of transaction costs. Although these subdivisions are not mutually exclusive his point, that each area has tended to develop with little regard to the others, is generally valid. Breton attempts to correct this, most notably the failure of the theory of democracy to take account of the existence and rôle of public goods. In tracing the development of the theory of bureaucracy as it has grown out of the Downsian theory of democracy a greater insight into the shortcomings and limitations of government oil policy is developed. This in turn can help to explain why government intervention in certain sectors does not achieve its desired objectives.

3.2 THE ECONOMIC THEORY OF DEMOCRACY

Downs attempts to construct an economic theory of representative government around the concept of the government as a monopoly. He relates the motives of rational, utilitarian individuals involved in the process of government to the political structure of a society, in this case a democracy. Within the framework of analysing the behaviour of rational individuals as citizens (voters) and in government (as politicians but implicitly as bureaucrats) Downs constructs two basic hypotheses about the operation of the political process in a democracy. From these hypotheses Downs is able to derive specific, testable propositions.

The central thesis adopted by Downs is that the activities of political parties in a democracy are analogous to those of entrepreneurs in a profit-seeking economy. Whilst entrepreneurs are often assumed to attempt to produce those goods which they consider will gain the most profits in order that their personal objectives and desires may be satisfied, the politicians adopt whatever policies they believe will gain the most votes for similar reasons.

Thus Downs' first major hypothesis is that political parties in a democracy formulate or adopt their policies so as to maximise their votes. As a result the behaviour of political parties in their response to changing economic conditions, and their actions in changing those conditions can be analysed in the same way as can the actions of other economic agents. The considerable importance of this analysis is thus self-evident when considering the extent of the government's impact in a mixed economy not only as a policy-maker but as an employer, producer, consumer, etc. To facilitate the analysis of the implications of this thesis Downs assumes that every citizen rationally attempts to maximise his utility income, including that portion of it derived from government activity; this is Downs' second major hypothesis. Because members of political parties are motivated by the intrinsic rewards of holding public office Downs argues that they will therefore formulate policies as a means to holding office rather than seeking office to carry out pre-determined policies.

According to Downs, the ultimate constraint on the activities of governments is the cost of information to the citizen and also to the government which itself is generally operating 'in a fog of uncertainty'.[4] On an individual level the citizen would be acting irrationally by acquiring information as the marginal cost of gaining the relevant information would greatly outweigh his marginal, but well-informed, benefit resulting from the infinitesimal effect of his single vote. There is also the further constraint of the competitive political process in a democracy between the incumbent and opposition parties.

The Downsian theory has had significant implications for the development of positive public choice theory (i.e. of non-market decision-making). Downs' 'enthusiasm' for the simple majority voting rule has led to criticism by Tullock[5] as it can lead to logrolling and the serving of small interest groups. Tullock[6] has consistently maintained that the Downsian theory does not adequately deal with either of these phenomena. In his 1959 article, Tullock uses an example of one hundred farmers each requiring repairs to his own road in order to illustrate an important shortcoming of the simple majority voting principle. With no vote-trading each road repair is defeated by 99 votes to 1. With logrolling each farmer pays more in taxes for the repairs of the other roads than he obtains for the repairs to his own road. This, according to Tullock, is due to the problem of 'revolving majorities'. Each individual farmer thinks he is paying one-hundredth of the cost of the total repairs and receiving one fifty-first of the total benefit. However, some of the other fifty farmers he has made a deal with, in order to reach a majority, will be offered deals by the 49 farmers originally left out of the first deal. Thus as many as 100 bills will be paid when each farmer is calculating from the basis of only 51 bills. Thus Tullock shows that the simple majority voting principle may lead, via a system of 'revolving majorities', to over-investment and the super-optimal size of government.

Downs[7] rejects the idea that the outcomes above are a consequence of logrolling and majority voting and maintains that Tullock's conclusions (that government activity which benefits minorities will receive disproportionate allocations of resources in comparison with the benefit to society as a whole; taxes of general impact will include exemptions favouring special groups and the government budget will be too large as individual voters will rationally support a level of government spending which is irrational for society as a whole) result from Tullock's *seriatim* assumption whereby voting takes place on a continuous stream of proposals. Downs' criticisms of Tullock's example of the one hundred farmers are that the *seriatim* assumption would not hold up in the real world and also that the tendency for the government to oversupply legislation giving a dis-proportionate benefit to special interest groups would be more than offset by the tendency to undersupply general interest legislation due to the free rider problem and the complementary lack of incentives for voters to become informed. Tullock has nevertheless been consistent in his preference for a two-thirds majority vote for import-ant issues even though each individual does not have an equally weighted vote.

A further important conclusion reached by Downs and derived from his party-motivation hypothesis is that democratic governments tend to favour producers more than consumers in their actions. Because of the strong common objectives of producers, combined with the costs of gathering information, a group of individuals, as producers, will be able to articulate clearly their policy preferences to the government who will themselves be keen for any indication of public preference and will thus pay more attention to the desires of producers than to consumers whose interests are more diverse and, due to the cost of information, more obscure. This idea can be extended to explain, at least partly, the importance of political pressure groups in general. Generally, pressure groups are relatively small, well-informed organisations with a very strong common con-viction on one aspect of government policy.

Buchanan and Tullock[8] approach the problems of logrolling and the optimal voting rule in some considerable detail. It is as part of this analysis that they come to promote the idea of 'reinforced majorities' for deciding major issues. The simple majority rule is generally not optimal according to a Pareto solution and it ignores strengths of preferences. Because the majority rule enables the individual in the dominant coalition to gain benefits from collective action without bearing the full marginal costs properly attributable to him, the simple majority rule will result in a relative overinvestment in the public sector if the standard Paretian criteria are accepted.

Logrolling may be explicit or implicit. With implicit logrolling a whole complex package of policies is offered, including some which would have only minority appeal. An individual then votes for the whole package on the basis of a strong preference for one or two items, despite opposition to other policies in the package.

Tullock[9] maintains that in virtually all real-world voting situations implicit logrolling takes place, and while this may have beneficial results it can also have defects.

Tullock clearly illustrates the benefits of logrolling[10] by using a simple example. In this similar example there is a society of twenty voters confronting a collection of policies to be paid for by a tax of £1 on each member of the society, and each individual will benefit exclusively from the policies. If one policy costs £20 to implement and only one individual accrues a benefit worth £35 the bill, without logrolling, would be defeated by 19 votes to 1, although it is preferable for the bill to be passed. With logrolling, assuming there are similar policies for other voters, it would be passed. However, it could be the case that logrolling has undesirable consequences. If, in the same system, the return to the individual is £15, the individual would be keen to trade votes on ten other issues which would thus net him £4 on the agreement. Thus a bill would be passed costing society £20 for a policy worth only £15. Tullock's 'radical' solution to this problem (although he acknowledges that there is no way to ensure that only those bills which benefit society get passed with logrolling) is to dispense with the simple majority rule and introduce re-enforced majorities.

Buchanan and Tullock's[11] analysis of various voting rules using 'game theory' gives rise to the idea of the existence of some optimal voting rule. By summing the external costs (which are the costs to individuals resulting from the passage of bills which are detrimental to individuals under different voting rules) and the bargaining costs (the more people needed to pass a bill the greater will be the costs of negotiating, i.e. investment of time and energy) the total cost incurred by society under various voting rules result. In Figure 3.1 the horizontal axis shows the voting rules that might be adopted, with a

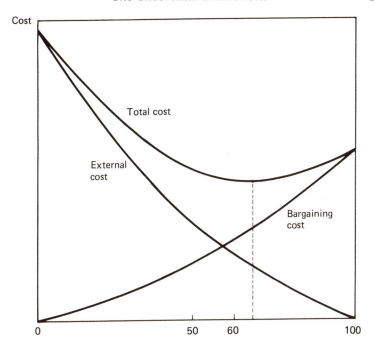

FIGURE 3.1   *The cost of democratic decisions*

SOURCE   G Tullock, *The Vote Motive* (London: IEA, 1976) p. 52.

one person dictatorship at the left and unanimity at the right. The total cost, therefore, can be derived with its lowest point being the optimal voting rule. The optimal voting rule will only by coincidence be at the 51 per cent simple majority level.

Further criticisms of Downs were outlined by Tullock[12] who maintains that although Downs's theory gives reasonable predictions for two parties it is not so good for more than two parties and also that most political choice decisions are not one dimensional. Tullock uses a similar analysis to that of Riker[13] to introduce intra-party rivalry as a constraint on the activities of a political party. Figure 3.2 employs the concept of 'issue space'[14] which has the obvious drawback of having voters' preferences equally scattered over the area portrayed. If one party mistakenly takes up position A, then the vote maximising reply is position B. All those voters whose optima lie to the left of line I will then vote for B, but for voters at the left end of the issue space this represents only a small gain. If the second party takes

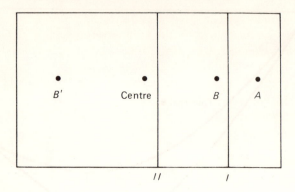

FIGURE 3.2   *Vote maximisation and intra-party revolution*

SOURCE   G Tullock, *Towards a Mathematics of Politics* (Ann Arbor: University of Michigan Press, 1962).

position B' it can still gain a majority and out of all those who had voted for B, about two-thirds of them will prefer B'. Thus as a result of intra-party rivalry B' would be adopted instead of B.

The switch from B to B' is a change from a position which gains a little for a large group of people to a position which ensures a larger gain for a smaller group. A party with a well-established organisation of professional politicians in power might choose the B strategy whilst an extremist wing, trying to take control of the party, would favour policy B'. This type of intra-party revolution could result in those party members favouring policy B being forced out of the party; a scenario not dissimilar to events in the Labour Party in the early 1980s.

Using a similar two dimensional diagram Tullock[15] is able to show that in a three party system the parties will not all tend to cluster around the centre (as they would in a two party system) but will tend to move away from the middle, making the policy differences between them greater than in the two party system. In a two party system with both parties trying to adopt a central position, the acquiring of information concerning voters' preferences is crucial as one party could be defeated even though it has adopted all but one of the policies of the opposing party. Generally the incumbent party is better able to gather information, for example, due to resources available in the Civil Service.

Breton,[16] like Downs, maintains that it is the actual institutional framework employed in a representative democracy that 'shields' politicians from knowledge of the preferences of citizens. This occurs due to three characteristics of the framework of democracy:

(1) democratic rules (i.e. voting rules);
(2) the length of the election period;
(3) the degree of full-line supply (i.e. how direct the voting system is).

Breton questions the validity of the base hypotheses from which Downs constructs his model. Breton argues that more and more media time is allocated to informing the public about political news and this is especially true in an election period when the cost of acquiring a minimal amount of political information to the individual is virtually zero. Breton is also dissatisfied with the constraints Downs puts on the activities of political parties. Whereas Riker and Tullock also considered the possibility of an intra-party revolt as a constraint, Breton maintains that the Downsian constraint of the competitive party system intuitively does not seem very strong when general elections only take place perhaps every four or five years. Breton therefore crucially introduces public and non-private goods (non-private goods are defined as goods which may not be available to everybody but have the property that the amount available to one individual does not reduce that available to others by an equal amount)[17] as a constraint on the behaviour of politicians in a way that is 'very similar to the constraint that a production function imposes on the maximisation of profits by entrepreneurs supplying private goods in competitive markets'.[18] Breton's method of analysis combines and develops much of the work of Buchanan and Tullock on the theory of decision rules and Downs' work on the theory of democracy, and introduces the theory of public goods and the rôle of bureaucracies in controlling and influencing the supply of public goods and policies into this framework.

On the demand side, Breton[19] defines 'the degree of coercion' as the difference between the 'amount' of public policies desired and the amount provided. The larger the difference the greater is the degree of coercion and, *ceteris paribus*, the greater is the likelihood that an individual will attempt to affect his own position by influencing politicians. It is as a result of government coercion and individuals' response to it that governments are informed of citizens' preferences and desires. The individual will only indulge in political participation if the coercion applied to him exceeds his personal 'coercion threshold'. The individual will then attempt to eliminate coercion by working for change in government expenditure policies and/or in fiscal policies. This process could result in political support for the Opposition, the formation of pressure groups, broader social movements, the private provision of non-private groups, migration, voting, etc. Although Downs did implicitly appreciate the rôle of political lobbyists Breton examines their behaviour in some detail.

Using the example of the United States Medical Association, Breton argues that people join pressure groups not only because of government coercion but because the pressure groups' intrinsic need to survive necessitates their encouraging people to join. A pressure group will then support or oppose some policy if, for a majority of their members, government coercion exceeds their coercion threshold. Thus a pressure group will be more effective the more homogeneous its members are in their personal preferences.

Although each citizen's coercion threshold is determined by himself or herself the actions taken by an individual independently in order to reduce government coercion could affect others. For instance if an individual 'votes with his feet' in response to government coercion and thus leaves the area, the tax burden on everyone remaining will increase pushing those at the margin over their coercion threshold forcing them into political participation of some sort. There would also be different thresholds for different forms of participation. For instance the threshold above which an individual is moved to sign a petition or enter into some relatively passive form of participation would be considerably lower than the threshold above which an individual decides to join an active pressure group or move home.

Political participation involves various costs, of organisation (i.e. costs of recruitment, bargaining and administration) and of communication (i.e. costs of propaganda and transport) and as these costs vary there will be a change in the kind of political instruments employed and of the intensity of their use.

The politician supplying government output will attempt to maximise his utility function:[20]

$$U_p' = U_p \ (T, a_m) \qquad (p = 1, \ldots P)$$
$$(m = 1, \ldots M)$$

where: p denotes a given politician, T denotes the probability of re-election, a_m denotes variables associated with the politician's personal ambitions, e.g. financial reward, kudos, etc.

This function will be maximised subject to some level of T below which that variable cannot be allowed to fall. For example, the rôle of the opposition party in Britain is not straightforward because, except at election times, the competition between the incumbent party and the opposition is limited. In the absence of effective competition the governing party will not do everything it can to satisfy the preferences of citizens as weak competition implies a low probability of defeat. A technical constraint (in addition to the physical constraints imposed by the production technology governing the combination of factors of production) on the maximisation of this utility function is the extent to which government coercion pushes individuals into political participation. Breton outlines four activities the government can employ by which to limit the coercion it imposes on citizens:

(1) discriminatory policies which can be tailored to small groups;
(2) discriminatorily enforced laws;
(3) indulgence in implicit logrolling;
(4) seeking to alter preferences of citizens.

3.3 NISKANEN'S THEORY OF BUREAUCRACY

With the introduction of the influence of supply of public goods on the behaviour of politicians, Breton considers the theory of bureaucracies as an integral part of his theory of representative government. Breton's recognition of the importance of the bureaucrat and the bureaucratic structure of the governmental system in the UK is a significant advance on previous works. Breton explicitly examinesx the conflict of interests between the politician (whose actions are already unlikely to reflect accurately the preferences of voters due to the numerous imperfections involved in any system of representative democracy) as the supplier of public policies, and the bureaucrat who will attempt to influence and shape the decisions made by the politicians in order to satisfy his own objectives.

When analysing the implications of government policy in the North Sea with regard to the economic efficiency of an industry it is important to be aware of the background in which that policy was formulated and implemented. It is therefore necessary to examine the economic theory of representative democracy in order to take account of the distortions and inconsistencies resulting out of the system of democratic government as they affect the politician in the government. It is also necessary to consider the economic theory of bureaucracy as the bureaucrat has a crucial relationship with the government and the politician and with industry.

The long-standing relationship a government department would have with an industry could result in the development of a 'special relationship' which the politician, because of the transient nature of the job, would not have. This can be simply illustrated by modifying an example used by Tullock.[22] A politician having to decide on a method of licence allocation for North Sea oilfields will be affected by the pressures and distortions imposed on him by the political process (such as from the majority rule, cycling, logrolling, pressure groups, etc.) which will affect the policy adoption process and the policy implementation process. The politician then has to employ a government department for research, information and the execution of the policy; and thus the bureaucrat will approach the subject with a view to protecting or furthering his personal objectives and will attempt to influence the politician for that purpose.

The power of the government bureaucracy stems from the characteristics of its relationship with its sponsor, in this case the government. Work by Downs[23] and Niskanen[24] form the basis of much of the work on the economic theory of bureaucracies and Niskanen's model is used in this chapter as the base point from which further contributions are introduced.

Niskanen[25] sees the relationship between the bureau and its sponsor as a bilateral monopoly. However, because there is generally a significant difference in the information available to the sponsor and to the bureau, the sponsor often cannot exploit its position as a monopsony due to lack of information. The sponsor is also not a profit seeking enterprise and officers within the sponsoring organisation (i.e. politicians in the government) have many functions to perform and are therefore essentially only part-time overseers and supervisors of the bureau's activities. The bureaucrat on the other hand often has a stronger relative incentive to obtain information (and is able to work full time in doing so) and is able to obscure or withhold information from the sponsor if it is in the bureaucrat's interests, as he sees them, to do so. Niskanen combines these factors to derive the crucial assumption of a passive sponsor which knows the budget it is prepared to grant for a given quantity of services but does not have the incentive or the opportunity to obtain information on the minimum budget necessary to supply the services.

As with Downs' economic theory of democracy Niskanen makes the assumption that bureaucrats behave in a rational utilitarian way analogous to a profit-maximising entrepreneur. On this assumption a bureaucrat will attempt to maximise the size of his budget. Although the personal efforts required to manage a bureau will be greater the larger the bureau is (or the larger the bureau's budget is), an increase in the budget will increase the effort required by a less than proportionate amount. The bureaucrat's instinct for survival will also lead him to maximise his budget as his personal work-load will lessen if he can provide increasing budgets for his subordinate bureaucrats to offer in salaries and contracts. Furthermore, the nature of the relationship between the sponsor and the bureau, in that the sponsor expects the bureau to expand and seek new programmes, often necessitates the bureau's seeking a larger budget.

In pursuit of this maximand (to maximise the budget) Breton[26] lists some of the techniques the bureaucrat may employ:

(1) overestimating benefits and underestimating costs of projects (not only through natural project enthusiasm but as a conscious policy);
(2) favouring a rate of discount in estimating the present value of benefits and costs which will make large projects appear more profitable than they really are;
(3) supporting the introduction of elaborate machinery to deal with rising costs;
(4) re-defining problems to keep them up to date;
(5) favouring the correction of 'wrong' policies rather than abolishing them;
(6) supporting projects that require transfers in kind rather than money as the former have a higher labour employment per unit;
(7) favouring economic planning.

The ultimate constraint on these activities and on the size of the bureau is that the total output of services that the sponsor expects the bureau to provide does not exceed a given budget. However the very nature of the output makes it difficult to quantify what precisely is expected. As the bureau exchanges its total service for a total budget the demand function facing a bureau is not a relation between price and quantity but between the marginal value of a service and the level of the service. Because the bureau is a monopoly, its marginal cost function is the marginal expenditure of the bureau at each level of service; it does not supply services at its marginal cost.

Thus Niskanen's model is based on two crucial assumptions – that bureaucrats maximise the size of their budgets and that they are monopolists who are able to impose their objectives on their sponsor. Using Niskanen's[27] notation, the potential budget available to the bureau during a defined period is shown by the budget-output function:

$$B = aQ - bQ^2 \qquad 0 \leqslant Q < \frac{a}{2b} \tag{3.1}$$

and the cost-output function is:

$$C = cQ + dQ^2 \qquad 0 \leqslant Q \tag{3.2}$$

with the constraint:

$$B \geqslant C$$

where B denotes budget, C denotes cost, Q denotes quantity.

The bureaucrat will attempt to maximise the expected approved budget subject to $B \geqslant C$. Thus putting equation (3.1) equal to equation (3.2) will give the cost-constrained output level:

$$
\begin{aligned}
aQ - bQ^2 &= cQ + dQ^2 \\
aQ - cQ &= dQ^2 + bQ^2 \\
Q(a-c) &= Q^2(d+b) \\
\frac{a-c}{d+b} &= \frac{Q^2}{Q} \\
\therefore Q^* &= \frac{a-c}{d+b}
\end{aligned}
$$

If the bureau is not constrained by $B \geqslant C$ there will be an upper output level:

$$
\begin{aligned}
B &= aQ - bQ^2 \\
0 &= a - 2bQ \\
2bQ &= a \\
Q_s &= a/2b
\end{aligned}
$$

The oversupply of bureaucratic output may be clearly seen diagrammatically in Figure 3.3. In a Pareto optimal position the bureaucrat would operate at output Qo where $B' = C'$. However, the bureaucrat seeks to operate at Q^* where area E equals area F and the bureau captures the equivalent of all the consumer surplus gains of output Qo. If the sponsor's 'demand' schedule is so far to the right, or inelastic, that the benefit of the bureau's output to the sponsor falls below zero before $F = E$ the bureaucrat then operates at output Qs, where the sponsor is satiated and the constraint that total budget equals or is greater than total costs does not apply.

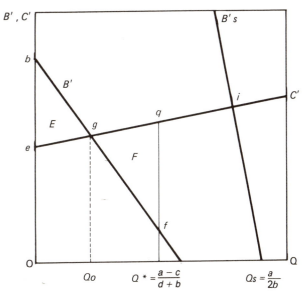

FIGURE 3.3   *The equilibrium output of a bureaucracy*

B' denotes the marginal benefit of the policy to the sponsor
C' denotes the marginal cost of the policy to the sponsor
$F = E$
$oeqQ^* = obfQ^*$
SOURCE   D. Mueller, *Public Choice* (Cambridge: Cambridge University Press, 1980),
p. 161.

Thus in Figure 3.3 where the bureau is a competitive purchaser of factors and provides a single service at low levels of demand (represented by the marginal valuation function, B') the equilibrium output of the bureau will be in the budget constrained region where $oeqQ^* = obfQ^*$. Niskanen[28] describes this bureau's output at

$$Q^* = \left(\frac{a - c}{d + b} \right)$$

as having no 'fat' as the total budget just covers the total costs. For the higher demand condition, $B's$, the constraint $B \geqslant C$ does not apply and the output of the bureau will only be constrained by demand. Thus output will be where the marginal value of the bureau's output (as perceived by the government, not the public) is zero, $Qs = a/2b$. Here the total budget is larger than the minimum total costs of that output (i.e. the area of the triangle formed by the $B's$ curve and the two axes is larger than the area of the polygon $oeiQs$). At this upper equilibrium level of output, Qs, there is 'fat' in the bureau, it will have no incentive to be efficient and will attempt to expand its expenditure in order to exhaust its approved budget.

Niskanen extends this basic model of bureaucratic output dropping some of his assumptions. If a bureau is able to exercise factor price discrimination both the budget and the output will tend to be higher than with no factor price discrimination. However, the upper equilibrium output level constrained only by demand is the same with or without factor price discrimination. This is due to the bureaucrat not exploiting the opportunity for factor price discrimination as it could not increase its budget by doing so and because the bureaucrat will not want to lose the support of those sponsors (government officers, politicians) who represent those factor interests. This is another area where pressure groups are important in influencing government output, here by applying pressure indirectly on the bureaucrat via the democratic voting process. Groups with a high relative demand for the output of a bureau and the owners of factors used in producing the output will capture the net benefits (equivalent to area E) and hence will be strong supporters of the bureau's higher than optimal output, Qo.

The most damaging criticism of Niskanen's model is based on the fundamental assumption of the bureaucrat attempting to maximise the size of his budget and being able to do so because of his characteristics as a monopolist. The politician (as the sponsor) is unlikely to be totally passive. Breton and Wintrobe[29] drop the assumption of the bureaucrat being a simple monopolist. They suggest[30] that the power of a bureau with respect to the sponsor is not

a result of the bureau being a monopoly but is more greatly dependent on its ability to distort and conceal information from its sponsor
and, closely related to this, on how technical and specialised that
information is. Breton and Wintrobe's second suggestion is that individual bureaucrats are not monopolists and in their consideration of
their career prospects they are aware of the sponsor's evaluation of their
performances. To an extent this conflicts with their first suggestion in
that the very nature of the bureau/sponsor relationship makes it very
difficult for the sponsor to assess the performance of the bureaucrat.
Mueller[31] backs up this criticism of Niskanen's bureaucratic maximand noting that objectives of the bureaucrat (for instance salary,
power, prestige, etc.) are not all necessarily positively and monotonically related to the size of the budget.

Thus the bargaining power of a bureau stems from its ability to
withhold information from the sponsor and also on the extent that
the information is highly technical noting that the politician is likely
to be a lay-person in that field. Breton and Wintrobe thus consider
the outcome of the sponsor being able to exert some kind of control
over the activities of the bureau. In Figure 3.3 the politician desires
output Qo at a budget cost of $oegQo$ and the bureau desires output Q^*
with a budget of $oeqQ^*$ (equals to $obfQ^*$). The difference between the
budget that the bureau succeeds in obtaining and that desired by
politicians is equivalent to the amount of consumer-surplus captured
by the bureau (area F = area E) and this is Breton and Wintrobe's
measure of 'control-loss' (defined as the cumulative discrepancy
between the actions of subordinates and the desires of superiors).
Breton and Wintrobe introduce the use of antidistortion devices
which will, at some cost, reduce the amount of control-loss; thus the
sponsor will employ the mechanism up to the point K (Figure 3.4),
where the marginal cost of the device equals the marginal benefit.
The area OHJ (Figure 3.4) is equivalent, in monetary terms, to area
E in Niskanen's diagram (Figure 3.3). At K, total control costs equal
$OLIK$ and the reduction in the budget (at output Q^*, Figure 3.3) is
$OHIK$. Now the optimum budget for the politician/sponsor given the
use of some control mechanism, is $oegQo$ (Figure 3.3) plus the
money equivalent of KIJ (Figure 3.4). Conversely, the optimum
budget can be seen to be the budget cost of output Q^* on Niskanen's
diagram minus the money equivalent of $OHIK$ on Breton and Wintrobe's diagram (Figure 3.4).

If the cost of the antidistortion device is prohibitive, and the
marginal cost of employing the control mechanism is always greater

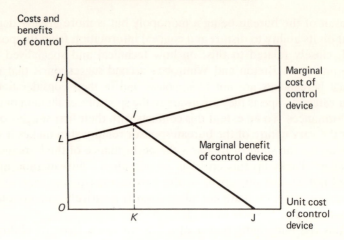

FIGURE 3.4　*The introduction of a control device*

SOURCE　A. Breton and R. Wintrobe, in *Journal of Political Economy*, February 1975, p. 200.

than the marginal benefit from its use, then the outcome is consistent with Niskanen's model and the bureau captures the consumer-surplus. The potential effectiveness of control devices depends on the source of inefficiency within the bureau. To eliminate the oversupply of output the sponsor needs to acquire information about the bureau's actual cost curve. If the inefficiency results from X-inefficiency the sponsor must have the information about the bureau's cost curve, plus an estimate of the true minimum cost of supplying the service which is thus a more costly operation.

There seem to be observable inconsistencies between 'actual' bureaucratic behaviour and the Niskanen model. It may be the case that the bureaucrat can further his career by reducing the size of his budget and such cases would tend to limit the application of the Niskanen model to certain types of bureaucracies. Breton and Wintrobe maintain[32] that more important is the tendency towards an automatic increase in government expenditure (and therefore the budget size of government bureaucracies) resulting from progressive taxation, economic growth and inflation. Breton seems here to contradict himself. If the behaviour of bureaucrats is examined within the overall theory of representative democracies this automatic increase in government revenue and expenditure should make no difference as there would come a point where a political party could gain office by adopting a policy of reducing public expenditure. Assuming a majority of the electorate do not consider this point (Breton's coercion threshold) has been reached public expenditure and the size of bureaucracies will continue to increase – but not automatically.

3.4 A MODEL OF THE SUPPLY OF PUBLIC OUTPUT

Breton[33] builds a model to determine the equilibrium supply of public output by combining both supply and demand conditions. On the demand side there is the democratic process and the effects of various voting methods and other distortions (costs of information, coercion threshholds, 'cycling' or the 'paradox of voting', logrolling, etc.) that may occur in this process which all contribute to affect policy formulation. On the supply side there are the constraints on the activities of politicians from political competition, intra-party rivalry, conflicts with bureaucrats and conflicts with the politician's own self-interest objectives. Breton combines these factors to construct a theory of bureaucratic supply, but by introducing a non-passive sponsor who is likely to employ some control mechanism the bureaucrat's maximand is no longer simply to maximise the size of his budget but more broadly to maximise his wealth (which is not necessarily positively and monotonically related to the budget size). If the politician is in a 'powerful' position the costs of political participation are low and more voters than are needed by the politician for election have made known their preference in his favour. Thus the output of public services will be determined by the influence and power the bureaucrat has over the politician and to what extent their personal objectives differ.

In Breton's diagram (Figure 3.5) as p^* (the cost of political participation) falls and more citizens are coerced into political activity Np (representing the number of people who vote) will rise. As Np increases over Nr (the number of votes needed by the politician for election) the politician becomes more 'powerful' in that he has a greater degree of freedom to adopt policies which may be contradictory to the desires of some sub-sets of his voting supporters as long as he satisfies a minimum of Nr. However, although the politician has the mandate of the voters to implement a certain policy, P, if P is not consistent with the policy, B, desired by the bureaucrat, bargaining will take place between the politician and the bureaucrat as to

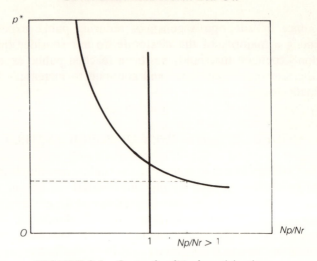

FIGURE 3.5   *Costs of political participation*

SOURCE   A. Breton, *The Economic Theory of Representative Government* (Chicago: Aldine, 1974)p. 181.

whether P or B or some compromise policy will be implemented.

For $Np/Nr < I$ the politician in power does not know which policies to adopt in order to be elected as not a sufficient number of people are using the various instruments of political action. Thus search activities will take place and the incumbent politician will probably be in possession of more information than the opposition candidate. The incumbent politician will attempt to lower the cost of political participation by lowering peoples' coercion thresholds and by lowering p^*, Np/Nr will rise.

To locate an equilibrium output of a bureaucracy one would have to integrate fully Breton's wealth maximising bureaucratic model into a theory which takes account of the demand for government output and the relationship between the politician and the bureaucrat. If, where $Np/Nr < I$, the politician's preferred policy, then P (assuming only one policy and not a complete package of policies), implies an output of public service of Qo on Niskanen's diagram (Figure 3.3), assuming cost constrained conditions. The bureaucrat would desire an alternative policy consistent with output at Q^* (assuming the bureau is dealing with only one policy and that policy could be equated to a budget level) or would attempt to increase the output implied by policy P to Q^*. With strong voter support the politician could attempt to employ control mechanisms so as to minimise control-loss. However, if the policy is not a government priority (the public backing for it is limited) and the technical information available to the bureau and not to the government is extensive, then the bargaining power of the bureaucrat *vis-à-vis* the politician would be considerable. The extent that the output of the bureau is more than Qo (and less than Q^*) when the government employs control devices depends on their bargaining strengths and then (assuming the bureaucrat is in a strong opposition) on whether the bureaucrat prefers to take the excess budget in the form of larger output or in the form of higher costs on the inframarginal units of output. This in turn will depend on the sources of the inefficiency and how effective the controls are in reducing the inefficiencies.

The degree to which control devices do work thus depends on the source of the bureau's inefficiency and the relative strengths of the bureaucrat's and politician's bargaining positions. The present government, with a large parliamentary majority and a high priority of reducing government expenditure has been seen to go to considerable lengths in attempting to reduce bureaucratic inefficiency within the public sector. The fact that governments do employ control instruments (i.e. they are perceived to be not too costly to justify their use) implies a further criticism of Niskanen's monopoly assumption.

In criticizing Niskanen's theory of bureaucracy, Breton and Wintrobe[34] argue that competition between bureaucrats within an organisation exists and is likely to prevent the oversupply of bureaucratic output. The nature of UK government bureaucracies is such that it is very difficult to measure both bureaucratic output and input. Within a bureaucracy efficiency and competence are likely to be judged according to the objectives of the bureaucracy as a whole. If the overall objective of the bureaucracy is for increases in budget size, then subordinate bureaucrats are likely to attempt to fulfil this objective. Moreover, it is not competition for jobs between bureaucrats that is a constraint on the maximisation of budget size but competition between bureaucracies for the supply of policy that is likely to be the constraint.

3.5 IMPLICATIONS FOR NORTH SEA OIL POLICY

For the purposes of this book it is necessary to consider the factors above which affect policy formulation and implementation. This involves both the Downsian type theory of democracy and Niskanen's theory of bureaucracy and their respective modifications and extensions. In analysing the performance of an industry in terms of economic efficiency using scenarios with differing assumptions as to the extent of government intervention, it is necessary to examine the process by which government intervention takes place and the effects of the numerous conflicts and imperfections which act to distort and confuse government action. This process is of especial importance when examining government policies purporting to increase economic efficiency or to protect the 'national interest'. Casual observation should reveal that if in order to achieve the economically efficient allocation of resources the political party would have to undertake a vote-losing action, that policy is unlikely to be implemented. Even if it is undertaken, by the time the policy is subjected to various pressures and influences within the democratic process it is likely that the final policy which is implemented will be significantly different from the one originally advocated. Thus it seems naïve to assume that as soon as an individual moves from organised market activities to some kind of government position he 'shifts his psychological and moral gears'[35] and becomes an altruist.

When discussing specific government actions, in this case with respect to North Sea oil, a distinction has to be made between two broad types of policy 'distortions'. First, there is the distortion between an election promise and the implemented policy which results from 'internal' political influences. For example, a popular political policy is to control production rates in the North Sea. Although the economic wisdom of such a move is open to considerable doubt successive governments have adopted this policy apparently because of its superficial attraction to voters. The experience of the present government has been that it has not felt it can always fulfil this election 'obligation' (although there have been some development delays, e.g. Clyde) because it would conflict with other priority government objectives. These 'internal' pressures would develop from various groups both within parliament and from outside. There are many groups who would like to see North Sea oil revenues used in a certain way (for instance, for job creation programmes, reducing the PSBR, investment in manufacturing industry, etc.) and in the case of this example all will be trying to influence the government not to reduce production levels and government revenue. Second, there arises the situation when the government, on attempting to implement a policy, meets a distortion effect from the government department through which the mechanics of the policy are worked out. It is in this area that the theory of bureaucracy can increase understanding of the governmental decision-making process (and is also relevant to the behaviour of the bureaucrat in a multinational oil company). If the policy diametrically opposes the interests of the bureaucrat, the bureaucrat will attempt to block its passage either before its adoption (by lobbying the government) or afterwards (by non co-operation). Alternatively, a policy could be 'used' by the bureaucrat in such a way as to satisfy the government and further the interests of the bureau. An example of this could be the government's desire for a regulatory agency to oversee oil industry activities in the North Sea being absorbed into the Department of Energy rather than by the establishment of a small, quasi-independent agency as proposed by Conservative Opposition Members in the BNOC debate (i.e The United Kingdom Oil Conservation Authority).[36]

Thus, when discussing the economic performance of an industry, (albeit an industry characterized by many imperfections) and contemplating the effects of varying degrees of government intervention it is important to take account of the political process of policy formulation and implementation which itself is highly imperfect.

4 Licensing Policy

4.1 INTRODUCTION

The following chapter attempts to examine the system of licence allocation adopted by the UK Government in 1964. The decision to employ a discretionary system in preference to some kind of competitive auction has significant implications with regards to the offshore oil industry and its relationship with the government. The rationale behind this decision is highlighted with the aid of the economic theories of politics and bureaucracies. In order to analyse the implications of the discretionary licensing system as it has worked in the UK, each licensing round is examined as are the stated government objectives and criteria concerning each licensing round.

The first section of this chapter will concentrate on a comparison of the discretionary allocation system and the competitive auction system. The economic and political consequences of each system are considerable and they can be seen to have largely determined the pattern of development of North Sea oil. The second and third sections of the chapter concentrate on the terms and conditions of each licence round. The size and timing of the licence rounds and the objectives of the various groups involved, either directly or indirectly, in the licensing process are examined and explained. To an extent these sections of the chapter augment Chapter 2 in that the early licensing rounds trace the development of the offshore oil industry. Government statements concerning licensing policy and the conditions imposed in the licensing system were the main indicators of government policy on North Sea oil throughout the 1960s. However, although the consequences of the discretionary licensing system influence most other aspects of North Sea oil policy in that the precedent of Civil Service control is set, the consequences of the discretionary system are most relevant to Chapter 7 on taxation policy.

The concluding section briefly compares different types of auctioning systems and attempts to sum the effects of the UK licensing system on the exploitation of North Sea oil.

4.2 AUCTION AND DISCRETIONARY ALLOCATION SYSTEMS

In a competitive auction each bidder gains by giving up more expected economic rent to the point where all the expected economic rent has been captured by the government. Although the extent of the economic rent cannot be known in advance, provided the bidding is truly competitive, the auction system will extract the expected rent. For the purposes of this analysis, economic rent is defined as the excess of the value of the resources above all relevant costs (including risk). The realised rent may be more or less than the expected rent but the possibility of either supernormal profits or a loss should tend to cancel each other out. Over time as bidders acquire information and expertise the difference between expected and realised economic rent would tend to diminish. Thus the auction system of allocating licences, by employing the price mechanism, enables the government to capture the maximum economic rent as well as ensuring economic efficiency in that the successful bidder will be the lowest cost bidder.[1]

In a discretionary allocation system licences are awarded on the basis of a set of criteria established by the government. These criteria may include political or bureaucratic considerations and may be discriminatorily enforced. The reasons why, in 1964, the Conservative Government chose to adopt a discretionary system in preference to an auction system, in spite of the latter's economic advantages, may be explained at least partly by the economic theories of politics and bureaucracies.

In 1964, prior to any indications as to the extent of oil reserves on the UKCS, the government's overwhelming priority with respect to the exploitation of North Sea resources, was for speed. The reasons for this objective are explained in Chapter 2. However, the method by which it was facilitated was by establishing an environment in which oil companies would be encouraged to commit vast resources to an unproved area. The discretionary system, by not extracting the rent from the oil companies, allowed the companies to operate in relatively favourable conditions. It is therefore not surprising that the oil companies favoured the discretionary system as it amounted to a tacit subsidy from the government in return for a rapid exploitation policy.[2] A further political ambition which was an important consideration to the government because of its vote-maximising potential was that of nationalism. As will be explained in this chapter, the government was keen to be publicly seen to discriminate in favour of UK oil companies and industry. Whereas the competitive auction system would grant a licence to the lowest cost producer, the discretionary system could be employed as a vote-capturing policy illustrating the government's concern for a 'national' resource and for domestic industry.

Public perception of government activities is an important policy determinant (see Chapter 3). The discretionary regime could be used to create the impression of the government being in control of the activities of multi-national oil companies and protecting the 'national interest'. In a discretionary system the government could impose conditions in order to maximise political support. A related point is that the discretionary system created the perception of the government aiding smaller independent oil companies who would have been out-bid in a competitive auction. The preceding argument seems to understate companies' abilities to obtain finance from the money markets or to enter joint venture projects. If large oil companies have greater funds for bidding they similarly would have greater funds for exploration and development; exploration would not be affected.[3]

Because of the nature of the UK system of democratic government (see Chapter 3), government bureaucrats have considerable power within the policy process. Government bureaucrats would have been strongly in favour of a discretionary allocation system for various behavioural reasons highlighted in Chapter 3. The administrative responsibility of the discretionary system is the concern of government bureaucrats. A discretionary system necessitates considerable work in appraising licence applications on behalf of the administrators of the system. The acquisition of information and knowledge is of crucial importance to the government bureaucrat and it has been contended[4] that in an auction system the lack of information would inhibit the proper functioning of the system. In an auction system the leasing agency would need a certain amount of information (e.g. to determine the size and timing of auctions) but competition between bidding companies would ensure government capture of the economic rent; information to the auctioning agency is not crucial. For the government bureaucrat to allocate licences according to various political criteria, the bureaucrat would need considerable information. Licence applicants are invited to submit work programmes and for these programmes (designed to indicate the prospective licensee's competence) to be analysed, the bureaucrat must possess highly specialised knowledge. With information on the offshore oil industry, which is crucial in order to implement the discretionary system, the bureaucrat fulfils important objectives. Technical knowledge enables the government bureaucrat to influence politicians and political parties because of the dependence of the politician on the bureaucrat for advice and guidance in a highly specialised area. Implementation of licensing policy requires considerable work, specialists have to be attracted into the Department of Energy[5] thus increasing the budget and influence of the department. Through the discretionary system the government bureaucrat retains considerable administrative power[6] which, because of his superior knowledge relative to the politician, may be employed to further his own interests.

Over time, as the system becomes more familiar to both the government bureaucrat and the oil companies, expertise and knowledge of the system is accumulated. Bureaucrats in the government and in oil companies would be strongly opposed to changing the system because of the threat to their positions and influence acquired over time. Politicians similarly are unlikely to change the system because of the implicit admission of adopting a 'wrong' policy in the first place.

The administrative powers bestowed on the Department of Energy by the discretionary system also enabled the government to establish a framework of control. The oil companies were answerable to the government in that if their performance was inconsistent with government attitudes and preferences the companies could lose the opportunity of a licence award in the next round. The oil companies had to prove their worthiness according to the criteria of politicians and civil servants.

Thus because the discretionary system transfers economic rent to the licensee, the government 'will attempt to "buy" something for the economic rent'[7] such as requiring a rapid rate of exploration (as compared to the rate determined by the market), requiring reserves to be sold to the government at a lower than market price (i.e. natural gas and the Gas Council)[8] or by increasing taxation. Each of these methods has been used by the UK Government in the North Sea. In the Eighth Licensing round, the Department of Energy considered the efforts of oil companies in involving UK organisations in research and development of offshore technology. Thus the government was also attempting to 'buy' research and development spending for the UK.

4.3 THE FIRST FOUR LICENSING ROUNDS

The first Licensing Round in 1964 allocated 53 licences covering 348 blocks, out of an offer of 960, to 23 licensees (see Table 4.1). The UK participation in the acreage was 30 per cent including 3 per cent to the public sector (the Gas Council).[9] The five criteria announced by the Conservative Minister of Power were basically designed to promote the rapid exploration and exploitation of any North Sea resources. Although this could not be done without the involvement of foreign, mainly US, oil companies, the two principal domestic oil interests, BP and Shell, were favoured in the distribution of the blocks. The first criterion was a general statement of intent 'to encourage the most rapid and thorough exploration and economical exploitation of petroleum resources on the continental shelf'.[10] The second criterion, 'that the applicant for a licence shall be incorporated in the UK and the profits of the operations shall be taxed here', had the specific purpose of enabling the full taxation of profits. The third criterion, 'where the applicant is a foreign owned concern, how far British oil companies receive equitable treatment in that country' may have excluded a few foreign firms but was probably more concerned with protecting British oil interests abroad in countries which had companies operating in the North Sea. The fourth criterion, 'we shall look at the programme of work of the applicant and also at the ability and resources to implement it', had the specific objective of precluding speculation in the selling of licences by excluding those who did not have the financial and technical ability to carry out an active programme of exploration and drilling.[11] When combined with the first criterion this gives rise to an implicit form of competitive bidding, (though not in the form of cash). The fifth criterion, 'the contribution the applicant has already made and is making towards the development of resources of our continental shelf and the fuel economy generally', was the criterion that most favoured established domestic oil companies as any British company could make some sort of case for itself as having made a contribution to the 'fuel economy generally'. Kenneth Dam[12] notes that the 30 per cent of the blocks that were awarded to British interests included a more than proportional share of the most sought after blocks.

The criteria are open to wide interpretation by those administering the licensing round. This increases the administrative power of the government bureaucrat as the conditions of licences may be interpreted to coincide with bureaucratic objectives. Coercion may be applied by the government bureaucracy in order to achieve its objectives. Because of the considerable amount of information the bureaucracy would need to collect in order to interpret and assess 'the programme of work of the applicant and also . . . the ability and resources' that are available to the oil company, the bureaucracy could in the longer term attempt to steer government policy. That these criteria are vague and open to considerable interpretation similarly coincides with political objectives of the various ministers. Policies may be altered or added to at later stages without contradicting earlier publicly stated intentions. Furthermore, the greater the discretionary powers bestowed on departmental ministers the greater is their personal influence and prestige. Vague and flexible policy intentions allow ministers to respond to political change whilst at the same time fulfilling the need for a policy issue to be publicly known.

During the mid-1960s North Sea oil policy was not the crucial political issue it subsequently became in the 1970s. Nevertheless, certain political and bureaucratic requirements were fulfilled through the policy process. Political and bureaucratic objectives can be seen to have been protected by government licensing policy. The domestic oil industry, as a political pressure group, can also be seen to have protected its interests (i.e the fifth criterion). The Treasury and Inland Revenue's interests are catered for in the second criterion, although this would be a normal procedure. To an extent the Foreign Office would be involved in the activities of domestic oil companies with overseas interests. Thus it is clear that at this very early stage in the development of North Sea operations the major interested parties were careful to protect their interests for the future. With respect to government bureaucrats the allocation of North Sea licences was an end in itself. Whether or not resources would be discovered government bureaucracies would still be required to gather information, advise politicians, negotiate with oil companies and administer policy. These activities would fulfil Niskanen-type bureaucratic objectives of maximising budget size and influence by employing some of the tactics outlined in Chapter 3.

Information as to voter preferences concerning a policy issue is crucial to the politician or political party faced with various policy options. In a two party government system if neither party is confident in gauging voter preferences there will be a tendency towards consensus. Because of the dangers of losing votes to the opposing party each party attempts to minimise this risk by formulating a policy which is closely aligned to the other party's policy (see Chapter 3). Licensing policy in the 1960s was not a major political issue, the tradition of government involvement in the economy had been firmly established in the UK since the war and with the exception of the US, all oil producing states employed some sort of discretionary licensing system. The cost to the individual voter to become informed on the licensing issue was prohibitive relative to the gains the voter would receive by acquiring information. The government, although unclear as to voter preferences, was not prepared to incur the costs of informing voters (i.e. pushing voters across their 'coercion threshold' – see Chapter 3) because North Sea oil was a low policy priority in terms of vote capturing potential. Thus the Labour Government, in announcing the Second Licensing Round in 1965, did not fundamentally change the terms of the licences but, in differentiating their policy, did add two new criteria.

First, the government would now consider the contribution applicants had made to the UK balance of payments. Secondly, and more significantly, the government intended to facilitate participation by public enterprises in the development and exploitation of the resources of the UKCS. This referred to the National Coal Board (NCB) and the Gas Council. It is difficult to ascertain the weighting given to each licensing condition and this secrecy protects government bureaucracies from criticism and from public accountability. However, in the Second Licensing Round in which 1102 blocks were offered and 37 licences granted covering 127 blocks, the public sector share rose from 3 per cent to 6 per cent (not including options and assignments – see Table 4.1). The NCB could only engage in offshore activities following the NCB (Additional Powers) Act of 1966 after which it was able to take up offers from Conoco, Gulf and Allied Chemical (who had been keen to acquire some public sector connections following the government's declared intentions of expanding public sector participation). The NCB was thus able to buy a 50 per cent interest in all Concoco's licences, 40 per cent in some of Gulf's and a carried interest in the Allied Chemical licence. The Gas Council increased its participation interests from 31 per cent to 50 per cent in the group made up of Amoco, Amerada and Texas Eastern. Thus total British interest was 37 per cent of the total area, from 30 per cent after the First Round.[13]

The major change in licensing policy was the greater emphasis on state participation in offshore activities. With respect to future licences granted in the Irish Sea it would be obligatory for applicants to provide for participation by the Gas Council or the NCB through direct partnerships, options or other acceptable arrangements. It was hoped that this would give the Gas Council more experience as an operator so it would have greater technical knowledge, greater control over gas supplies and would be able to keep abreast of the latest developments enabling it to be more active on the UKCS. The rationale behind this decision was that in future licensing rounds added preference would be given to groups involving the Gas Council, the NCB and other British interests. This, to an extent, foreshadowed what the Labour Party intended to do in future rounds. In effect, the Gas Council and the NCB were given an option to take a share in licences after they had been granted. The Labour Government's actions after the Second Round are an important indication of their overall attitude to the North Sea. Two important political objectives were fulfilled by the Labour Government in the Second Licensing Round. Firstly, although the terms of the licences remained basically very similar to those of the previous Conservative Government, the two new conditions were intended to show the electorate that Labour's policy was different from its predecessors. This objective was achieved by the second new criterion (referring to state participation) which had the effect of introducing a traditionally socialist objective into offshore oil policy. The second political objective achieved by the Labour Government is related to, but distinct from, the first objective. By introducing the participation criterion the Labour Government was pre-empting criticism from within the Labour Party as to the socialist nature of licensing policy.

In the Third Licensing Round in June 1970, 37 licences were granted covering 106 blocks out of 157 offered. Although the total UK participation had remained about the same at 36 per cent (from 37 per cent in the Second Round) the public sector share was now 13 per cent.[14] Thus a significant trend had been established in the first three rounds of maintaining the UK share in licences and increasing the public sector share. The Third Round, instigated by the 1969 Labour Government, again increased the emphasis given to public sector participation. The Labour Party's emphasis on participation can partly be explained by intra-party rivalry. A recommendation in 1967 (by the Labour Party National Executive Committee) suggested the formation of a National Hydrocarbons Corporation to control production of oil and gas in unlicensed offshore blocks. The actual policy outcome was a decision to allocate licences to a newly created Hydrocarbons subsidiary of the Gas Council. This illustrates political bargaining within the Labour Party and concessions made to the left wing of the Party. There was also a degree of implicit competitive bidding in the Labour Government's licensing policy. Because licence applicants for Irish Sea acreage would only be considered if provision was made for participation by either the Gas Council or the NCB, private groups could be expected to offer these nationalised industries very favourable terms in order to secure a licence.

Subsequent to the Second Licensing Round the Labour Government instituted a review of licensing policy in order to consider the various methods of licensing available.[15] Although the final policy outcome was similar to previous licensing (except with added emphasis on participation by nationalised industries) the policy review did fulfil certain bureaucratic objectives. The updating and appraisal of policy is a desirable activity in itself with respect to government bureaucracies. Information is acquired and, as the workload increases, budget size also tends to increase. The government bureaucrat, due to the imbalance of information, is able to influence the politician with respect to the size and timing of licensing rounds. By employing a discretionary policy arbitrary powers of ministers and senior Civil Servants may result in decisions being taken on political or bureaucratic criteria which are detrimental to the offshore industry. In the mid-1960s there was a five year period between the Second and Third Licensing Rounds. This may partly be explained by the desire to acquire geological knowledge of the UKCS and the political desire to encourage the domestic offshore supplies industry. A further factor is bureaucratic control. Licensing rounds at regular intervals and of predetermined size could reduce uncertainty with respect to oil company planning. However, this would detract from the government bureaucracy's position as the overseer and protector of offshore oil and gas reserves on behalf of the government and of the country. The government bureaucracy's interest may be satisfied by the public perception of the bureaucracy being solely able to control policy implementation. Furthermore, politicians would consider the size and timing of licensing rounds in a political vote-maximising context. The first two rounds in consecutive years resulted in 90 licences being issued. This was a sizeable licence issue and largely accounted for the delay between the Second and Third Rounds. During the 1960s licensing policy was used by politicians in order to implement their overall rapid exploitation policy and as in the late 1970s when licensing policy was used to slow down development (relative to the market determined rate) these are policy decisions based on political rather than economic judgement.

Following the awards of licences in the Third Licensing Round the government perceived a change in factors affecting North Sea oil policy. Oil had been discovered by Shell, Phillips and BP in areas leased in the second and third rounds. These finds required further appraisal work in order to determine the extent of reserves. The Ekofisk oil field on the Norwegian Continental Shelf was discovered in November 1969 and by mid-1970 evidence 'suggested the presence of a significant oil producing basin'.[16] A further factor was the Libyan Government's success in increasing its take from oil company concessionaires in September 1970. At the same time there was concern at the slowdown in drilling activity in the UK sector and acreage surrendered as a result of the first and second licensing rounds could now be offered again. These factors encouraged the government to institute a Fourth Licensing Round in 1971 and the main policy criteria were, as in previous rounds, speed of exploration and development and ensuring the representation of domestic interests.[17]

The government decided to continue to use the discretionary method of allocating licences. This decision was taken largely on the basis of the past, perceived success using the discretionary method in terms of the speed of development and the increasing British interests that had come about in both the public and private sectors. The government believed[18] that because they wanted to give added impetus to North Sea activity an auction would not have been satisfactory as one third of the area in the Fourth Round was unexplored and the less attractive acreage would be neglected. It was also felt that because so many conditions and provisos would have to be integrated into the auction it would be unlikely to determine the market value of the territory.[19] More recently, it has been suggested[20] that the drawbacks of changing systems 'mid-stream' and wasting cumulative experience of past discovery regulatory procedure would be detrimental to the government and the industry. Also US companies would have had a distinct advantage because of their experience of competitive tenders. A further factor affecting the government's decision to continue with the discretionary licensing system was that it was felt that the benefits from the rapid development of the UKCS and the reduction of UK dependence on foreign oil outweighed the likely immediate financial benefits of an auction.[21]

In terms of the economic theories of bureaucracies, however, a significant reason for not changing to an auction method of licence allocation would have been opposition from government departments.[22] The Civil Service's desire for control and ability to 'steer' government policy would make it unlikely that such a policy change would occur. Nevertheless, the Fourth Round included an auction experiment. This was designed to acquire experience of the functioning of an auction and the acreage chosen was intentionally a range of attractive and less attractive areas. Information on the workings of an auction would have been desirable for the government bureaucracy in order that it could maintain flexibility and knowledge of policy alternatives. Because of official opposition to a complete change of policy to a competitive auction system it was unlikely that, however successful the experimental auction had been, it would have replaced the discretionary system.

In the Fourth Licensing Round blocks were offered in four areas:

(1) the Southern Basin of the North Sea;
(2) the Northern Basin of the North Sea; a few exploratory wells had been drilled and there was some optimism as to the possibility of finding oil;
(3) the Western Approaches to the Celtic Sea;
(4) the area to the west of the Orkneys and Shetlands.

Both areas (3) and (4) were largely untested. Under the discretionary system 421 blocks were offered and 267 were licensed; with a British participation of 44 per cent. In addition, under the experimental auction system, 15 blocks were offered, all of which were licensed; with a British participation of 22 per cent. Total British participation had thus increased to 43 per cent of the total area. The auction was a financial success raising a total of £37m. Of the 15 blocks auctioned, nine blocks fetched less than £1m, 5 fetched £51 000 or less and the lowest successful bid was £3200. The highest bid was made by Shell/Esso who paid £21m for a 'golden' block. That this bid by Shell/Esso was £13m higher than the next bid reflected the lack of geological knowledge at the time.[23]

A significant consequence of the auction experiment was that because the discretionary method failed to capture the economic rent for the government some system of recouping the rent had to be devised. Government attention turned to this issue. The development of a North Sea taxation system designed to collect the rent from oil is examined in Chapter 7.

4.4 1976 TO THE PRESENT DAY

The Fifth Licensing Round was announced in August 1976, five years after the announcement of the Fourth Round. This delay was largely a result of the considerable changes that had occurred both in the domestic oil market and in the world oil market. These changes, i.e. the oil price increases of 1973–4, perceptions of shortage, increasing discoveries on the UKCS, combined to politicise North Sea oil. In order to maximise the vote-capturing potential of North Sea oil political parties were concerned with developing major oil policies (see Chapter 3). The Labour Government outlined its policy intentions with respect to North Sea oil in a White Paper[24] in 1974. The White Paper announced an important new policy consideration which significantly altered the criteria by which licences were granted. It was intended that in future licences 'the licences shall, if the Government so require, grant majority participation to the State in all fields discovered under those licences'.[25] As discussed in Chapter 5, the original intentions of the government for participation to be based on some sort of 'carried interest' system were unacceptable to the oil companies. The 'carried interest' system enables the state to 'carry-over' its interests in a field during the exploration stages of an oilfield project. If the field is then considered commercially viable the government can exercise its option to participate. The percentage of the costs paid by the government can vary greatly between fields depending on individual participation agreements. The government is therefore 'carried through' the high risk stage and does not commit capital to ventures which turn out to be non-commercial. This initial policy objective was more likely to be a negotiating tactic by the government to be used as a base from which bargaining would take place.

The 1974 White Paper also announced the government's intention to establish a state oil company (BNOC) 'through which the Government will exercise their participation rights'.[26] The establishment of BNOC was the crucial element in the Labour Government's oil

policy. Ideologically it fulfilled a traditional Labour objective for
state intervention in key industries and it clearly differentiated Labour oil policy from Conservative policies. BNOC, under the ste-
wardship of Lord Kearton, was responsible for the participation
negotiations with oil companies operating in the North Sea and had
the backing of the Department of Energy, keen to expand its control
and the Government, keen for a discernible policy to be implemented. The Government's bargaining position was strengthened
by popular support for the policy. Frustration on behalf of voters
faced with higher oil prices and the perception of 'windfall' profits
accruing to multi-national oil companies enabled the Government to
take political action.

Although the Government was in a relatively strong bargaining
position it was nevertheless heavily dependent on the oil companies
(even with BNOC in existence) for the exploitation of UK North Sea
resources. By applying too much pressure on the oil companies to
reach participation agreements with BNOC the Government was
risking the oil companies' departure from the North Sea. Whilst these
negotiations were taking place there was concern over activity in the
UK North Sea slowing down.[27] Between 1974 and 1976 few new
production projects were announced and platform construction facilities in Scotland were facing shut-down because of the dearth of new
orders. Thus there were constraints on the Government's ability to
force oil companies into participation agreements and this difficulty
largely accounted for the lengthy participation negotiations (some of
which were still continuing in 1979 and 1980). The protracted participation negotiations largely explain the delay in the announcement of
the Fifth Licensing Round. Many large oil companies had not reached
satisfactory agreements and the government could not afford to
prohibit them from further licence awards; at the same time the
government did want to make it clear that they were determined to
finalise suitable participation agreements. Perhaps the most serious
confrontation between the Government and the oil companies was
with Amoco whose refusal to enter a 'voluntary' participation agreement resulted in their being totally excluded from the Fifth Round
allocation of licences.

The Fifth Licensing Round, when eventually announced, was of
relatively small size. In 1972, 282 blocks were awarded whereas in
1976 only 71 blocks were offered and licences issued for 44 blocks
(see Table 4.1). A further factor announced by the Government
concerning the size and frequency of licensing rounds was that

because of the 'violent fluctuations in work-load for the offshore supply industry'[28] following the Fourth Licensing Round the future licensing rounds would be more frequent, regular and smaller than they had been previously.

The crucial change in the conditions attached to licence awards in the Fifth Round was that BNOC (or BGC) received a 51 per cent majority interest in all blocks. The results of the Fifth Round were announced in February 1977 and BNOC was the operator on four blocks.[29] The Labour Government changed some of the terms of licences, again with the objective of differentiating Labour's oil policies and justifying the establishment of BNOC. Labour Ministers were critical of past Conservative licensing policies; as Mr Varley said, "A licence may last for forty-six years. Exploration programmes are required for the first six years. A licensee then surrenders half the territory – he chooses which half – but on the area he keeps he need do no more exploration for the remaining forty years".[30] However, Mr Varley's statement seems to ignore the market signals that the licensee will be continually receiving. If the licensee considers that the expected price of oil will be such as to justify further exploration, development and production, then it is unlikely that an area will be neglected for forty years. The new surrender provision seems to have been designed to encourage licensees to discover as quickly as possible the most promising areas of the licence area.[31] Under the Fifth Round, the initial licence period is for four years with an option for the whole area of a further three years and for up to one-third on the area for thirty years. Rental payments were also increased for the Fifth Round and the basis from which royalty calculations were made was changed. From the Fifth Round the $12\frac{1}{2}$ per cent royalty was levied on the tax value of the crude oil rather than on the (lower) wellhead value as in previous rounds which had allowed a deduction for the costs of conveying and treating petroleum.[32]

Thus the Fifth Licensing Round clearly shows how aspects of the economics of politics have affected the development of licensing policy and participation. The bargaining process between the Government and the oil companies illustrates how a changing political environment may alter the Government's ability to implement a policy. With respect to participation, altered political considerations were an important factor in the development of the policy before the bargaining process determined its final form. Popular support for government action, because of the growing importance of energy issues and the extensive coverage in the media, strengthened the

TABLE 4.1 *UK Offshore licensing rounds 1964–83*

Round	Date	Blocks* offered Area	Blocks* offered Number	Applications received No. Received	Applications received No. of Blocks Applied for	Applications received No. of Companies	Licences awarded No. of Blocks	Licences awarded No. of Licences	Licences awarded No. of Companies
First	1964	N. Sea	960	31	394	61	348	53	51
Second	1965	N. Sea Irish Sea English Channel	1102	21	127	54	127	37	44
Third	1970	N. Sea Irish Sea Orkney/Shetland Basin	157	34	117	54	106	37	61
Fourth (i) Discretionary	1971–2	N. Sea Irish Sea Celtic Sea	421	92	271	228)	282	118	213
(ii) Auction		N. Sea	15	31	15	73)			
Fifth	1976–7	N. Sea; Irish Sea Celtic Sea Orkney/Shetland Basin English Channel/ S.W. Approaches West of Scotland	71	53	51	133	44	28	64
Sixth	1978–9	N. Sea W. Shetland Basin							

Round	Year	Regions							
		Cardigan Bay/Bristol Channel; S.W. Approaches	46	55	46	94	42	26	59
Seventh	1980–1	N. Sea; W. Shetland Basin; Orkney/Shetland Basin; English Channel; S.W. Approaches	Specified area of Northern N. Sea; 80 elsewhere	125	97	204	90	90	157
Eighth (i) Discretionary	1982–3	N. Sea; W. Orkney Basin; E. Shetland Basin; Unst; Fair Isle; Forth Approaches	169	40	76	94	70	49	38
(ii) Auction		Bristol Channel	15	20	8	47			
Ninth (i) Discretionary	1984–5	W. Shetland Basin; Rockall Trough; Faeroes Trough; Morecombe Bay; Celtic Sea; English Channel; N. North Sea; S. North Sea; Central North Sea	180	117	107	134	80	74	
(ii) Auction		N. Sea	15	32	13	52	13	13	28

* Each Block is approximately 250 sq. kilometers

SOURCES   Department of Energy, *Development of the Oil and Gas Resources of the United Kingdom 1985* (London: HMSO, 1985) Appendix I; *Financial Times*, 24 May 1985, p.8.

Government's position in participation negotiations and the original decision to adopt a discretionary licensing system enabled the Government to incorporate the new conditions of licences into the system.

In the Sixth Licensing Round announced in August 1978, oil companies wishing to apply for licences would be given the 'opportunity' to offer BNOC more than 51 per cent equity in their blocks. Also companies could 'offer' to carry BNOC for all or part of its exploration and appraisal costs. Thus any operator expecting to receive additional UKCS acreage from the government would have to give BNOC more than 51 per cent equity and reduced exploration costs. The discretionary licence system also allowed the Department of Energy to grant BNOC licences outside the licensing rounds, again illustrating the Government's determination to support BNOC and the Department of Energy's willingness to implement policies under circumstances when its influence and power would be expanded.

Further conditions introduced in the Sixth Round were to consider the efforts of applicants in improving training in offshore work with the aim of giving more skilled jobs to UK citizens.[33] The Government attempted to aid small companies in the North Sea by stating that the operator for a licence at the exploration stage would not necessarily be the same as the operator at the development stage. This would make it possible for small companies without the financial resources to undertake development, to play a more important part in exploration and then farm-out their interests.[34] Increased regulation of licence issue increased the power of bureaucrats in the Department of Energy and at the same time coincides with the overall political ideals of the Labour Party then in Government. Policies tending to be nationalistic or understating economic efficiency (and the ability of small companies to raise capital on the financial markets) were politically popular and received support from government bureaucrats.

With the Sixth Licensing Round, as with other aspects of domestic oil policy, the Government significantly tightened its control of North Sea activities. Oil companies implicitly had to compete for blocks by giving BNOC equity interests of more than 51 per cent, by giving BNOC options to purchase or sell oil and by 'carrying' BNOC's interest.[35]

For the first time there were applications for all blocks available but Exxon made no applications at all and Shell, Chevron and

Conoco all restricted their bids, presumably to make clear their disapproval of UK offshore oil policy.[36] The Department of Energy's attitude seemed to be that as long as companies continued to apply for licences it would be possible to tighten up the terms yet further. However, Civil Servants in the Department of Energy were conscious of pressure from United Kingdom Offshore Operators Association (UKOOA) who were stressing the need for a milder tax and regulatory system.[37] Expectations in the offshore industry of a relaxation of government control were resting on a sympathetic Conservative Government entering office in 1979.

The proposals for a Seventh Licensing Round were announced in December 1979 and the Conservative Government formally invited licence applications in May 1980. The Conservative Government achieved electoral success at least partly because of its non-interventionist policy proposals. In the Seventh Round the government was keen to apply this overall policy to North Sea licensing. BNOC and BGC were to have the same opportunities to apply for licences as the private sector companies, but they would not have a mandatory majority interest in future licences.[38] A further significant change in the Seventh Licensing Round was that in an area defined by the Government, companies were able to apply for blocks of their own selection. Successful applicants for these areas were required to make an initial payment of £5m. The Government again changed the terms of licence surrender so that at the end of six years the licensee could retain half of the area originally licensed. Although BNOC no longer received a 51 per cent equity share in licences, awards were still conditional on prospective licensees agreeing to give BNOC the option to take 'up to 51 per cent of each co-licensee's share of any petroleum produced from the licensed area'[39] at the market price.

In response to the Seventh Round, 125 applications for licences were received, which was the largest number received by the Government since licensing began in 1964.[40] Forty-two blocks were awarded in company-selected areas and another 48 in other areas, thus raising an initial payment of £210m in total.

The Eighth Round announced in September 1982 was intended to boost activity in the North Sea. It was received by an oil industry highly critical of the offshore tax system. Immediate reactions in the industry were concerned with the likely impact on future activity in the North Sea of any continuation of the existing tax system.[41] Thus the Eighth Round afforded oil companies an opportunity to lobby the

Government and to influence public opinion concerning their dissatisfaction with the offshore tax régime. The bargaining process between government and industry continued and licensing rounds are an important means by which this process can take place.

A feature of the Eighth Round was the offer of 15 blocks for auction. Of the remaining 169 blocks, 131 were in 'frontier' areas and 35 in the Southern Gas Basin. The blocks auctioned were in mature oil areas of the central North Sea; thus the Government expected a good response in terms of revenue raised for the Treasury preparing an election-year budget. The decision to auction a selected number of blocks may thus have been based on short-term considerations with the Treasury, a bureaucracy with its own objectives.

In the Fifth Round the Labour Government announced its intentions for regular licensing rounds in both timing and size. The Seventh Round signified the end of this policy with respect to the regular size of rounds. It is evident that the licensing rounds are sensitive to political change which affects the objectives, priorities and rationale of the discretionary allocation of licences. Thus uncertainties accrue from the governmental rôle in the discretionary system which are in addition to the inherent uncertainties in the oil industry.

The results of the Eighth Round were announced in May 1983, although the outcome of the auction part of the Licensing Round had been made known earlier. In the Eighth Round the Government auctioned seven blocks out of the 15 on offer, raising over £32m (thus providing useful short-term revenue for the Treasury). In addition, licences were awarded by the discretionary system for 63 blocks out of 169 on offer. Thus a total of 70 blocks had been awarded in the Eighth Round, which was below the government's aim for between 80 and 85 blocks to be allocated.

The Ninth Licensing Round again included an auction, this time raising £121m. The two main features of the discretionary awards announced in May 1985 were as follows. Firstly, smaller independent companies were given preference in acquiring acreage in the more mature, proven areas of the UKCS. Secondly, large multi-national companies were directed to deep-water frontier areas and given the incentive of slightly more favourable surrender terms. A total of 93 blocks were awarded in the Ninth Round, exceeding government expectations.

4.5 THE AUCTION DECISION

The preceding analysis illustrates the extent to which the government may attempt to use licensing policy in order to control the offshore industry. The decision to adopt, in 1964, a discretionary licence allocation system conferred considerable administrative powers on government bureaucrats. Uncertainty with respect to the size and timing of licensing rounds can be partly explained by successive governments' willingness to use the licensing system to regulate the exploitation of North Sea oil. Licensing is seen as an integral part of governments' overall North Sea policy objectives. The rapid exploitation policy of the 1960s was encouraged by the licensing system which transferred the prospective economic rent to the licensees.

In the 1970s the politicisation of the oil market encouraged the Labour Government to extend the influence of the state into the offshore industry, in line with Labour Party ideology. The licensing system was used as a bargaining tool to persuade oil companies to enter into participation agreements with the newly established state oil corporation. The delay in the announcement of the Fifth Round and its relatively small size may be attributed, at least partly, to the lengthy participation negotiations. The Sixth Round was one aspect of an overall government policy showing a continuing trend of tighter control. The Seventh Round, instigated by a Conservative Party in office illustrated a 'ratchet effect' of government policy whereby the previous government's policies of intervention began to be reversed; but not to their initial states.

Preferential treatment given to British interests, and most notably to BNOC (up to the Eighth Round) is a constant theme throughout the licensing rounds. This is essentially a judgement taken on political criteria.

Licensing rounds have been used for strategic bargaining by the government and the oil companies have similarly responded by using the system as a bargaining tool. Dissatisfaction with the North Sea tax régime resulted in exploration and development delays and the reluctance of oil companies to accomodate Government participation plans has resulted in some companies not taking part in licensing rounds. Bureaucrats in government and in oil companies have accumulated expertise and experience and thus can manipulate the system to protect and further their interests. A radical change in licensing policy has therefore been consistently opposed by industry and government.

Opposition from BRINDEX and UKOOA[42] to an auction system is predictable on the basis of the economics of politics and bureaucracy. As stated, arguments that independent oil companies could not compete with the large multi-national oil companies fail to consider the ability of companies to enter joint-venture operations[43] and to raise capital on the money markets.[44] That auctions add to the financial problems of front-end loading is similarly doubtful. A royalty bidding scheme could spread payments over a project's lifetime and cash bids may be paid over time in instalments.[45] However, percentage royalty bidding may have a disincentive effect on the development of high-cost marginal fields and may cause the premature abandonment of fields because royalties are a tax on production.[46]

The problem of collusion in an oligopolistic market is relevant to licence bidding in the North Sea. Evidence, however, is 'consistent with the hypothesis that the market (for offshore oil leases) is highly competitive'.[47] This would seem to be affirmed by the fact that in the Seventh Round, 204 companies were involved in 125 licence applications and licences were awarded to consortia comprising 157 companies (see Table 4.1).

Perhaps the overwhelming problem associated with an auction system of licensing in one which accrues from the UK system of representative government. The world oil market is characterised by periodic shocks and considerable political and economic uncertainty. With a system of bonus (cash) bidding, in the event of an oil price increase similar to those experienced in the 1970s, public perceptions of oil company 'windfall' profits could encourage political parties to attempt to capture votes by introducing new or additional oil tax legislation.[48] To an extent this was the case in the US with the 1980 'Windfall Profit Tax on Crude Oil'. Thus with a system of licence auctioning uncertainty may remain due to politicians (and political parties) attempting to maximise their utility functions within the system of governmental democracy. As explained in Chapter 3, it is crucial for political parties to develop policies concerning issues which receive considerable media attention and political parties in power must be seen to be taking action. The risk of failing to maximise the vote-capturing potential of a policy thus often necessitates the implementation of policies largely for political rather than economic reasons.

Thus the institutional structure of government in the UK may cause uncertainties in an auction system of licensing. However, these uncertainties are considerably less than those present in a discretionary system. The main reason for this is the extensive power and influence of bureaucrats in a discretionary system. A bidding system would be administered by government bureaucrats but their functions would be limited to clearly defined areas and would not add significantly to uncertainties already present in the oil policy process.

5 Participation

5.1 INTRODUCTION

The purpose of this chapter is to examine the UK government's attempts to influence the domestic oil industry through the creation of a national oil company (NOC), the British National Oil Corporation (BNOC). Employing the methodology of Downs, Niskanen and Breton, the debate leading to the establishment of BNOC is analysed with specific reference to the bureaucratic and political pressures which have helped to determine the rôle of BNOC. The second section considers the function of BNOC after 1976 when its existence altered the framework of decision-making in the North Sea. The election of a Conservative Government in 1979 resulted in considerable speculation as to the future of BNOC; the final section of this chapter attempts to examine the conflicts of interest arising out of this change of administration and the implications with respect to the future rôle of BNOC. Thus the structure of this chapter tends to follow the stages of the government policy process from the perception of a political need for a policy to the implementation of that policy. The various factors affecting the policy process are not necessarily exclusive to each stage and each stage does not, of course, have a beginning and end which can be pinpointed to an exact time or event.

The economic theories of politics and bureaucracies (see Chapter 3) facilitate the analysis of the various forces involved in the government policy-making process. By applying these theories to the government/BNOC relationship an insight into the environment in which government policies evolve can be gained. It is hoped that within this framework predictions of future trends pertaining to the rôle of BNOC can be forecast. The overall objective of this chapter is to identify the forces and pressures which are inherent in the government policy process so as to examine the likely outcome of attempts to intervene in the oil market.

The decision taken in the early 1960s to employ a discretionary rather than competitive bidding form of licensing established an important precedent with regard to the government's rôle in the North Sea (see Chapter 4). The discretionary system allowed the government[1] to put direct pressure on oil companies more effectively than an auction system would have done. In effect, oil companies would be answerable to the government in that if their performance was not 'good enough' they could lose the opportunity of a licence in the next round; companies had to prove their worthiness to the satisfaction of the government. The government was careful to create an institutional framework whereby it could at some period in the future intervene directly or indirectly in the North Sea. However, it is important to note that the transfer of economic rent to the oil companies, which took place because of the nature of the discretionary allocation system, was being recaptured through the tax régime (see Chapter 7). Thus state participation via BNOC was overwhelmingly concerned with control and achieving a greater degree of central management in the economy.

Two vital prerequisites for effective government control in the North Sea are, first an overall picture of the desired energy scene and a concept of the specific problem at hand; and secondly, expertise and information within the government itself are necessary. The 'overall picture' implies that the government should have an energy policy and within that policy BNOC should have a definite function to perform. Even if the principle of an energy policy is desirable[2] or feasible there is the added problem caused by the democratic system of government whereby it is quite possible that at least every five years a new government will institute a fundamentally different policy from the one already in existence. Moreover, the importance of the government possessing expertise and information superior to that of BNOC (the organisation to be controlled) is that otherwise there is a real risk of the rôles being reversed and BNOC steering the government (according to Grayson[3] this has happened in France and Italy with respect to SNEA and ENI). Again, even if the government does possess superior knowledge it must then have the ability to implement measures which will combat the numerous and complex problems and disturbances thrown up by the controlled organisation (BNOC). There is a further and crucial aspect to this relationship between the controllor and the controllee which is the objectives of BNOC itself and BNOC's ability to withhold and select information if it is in the Corporation's interests (as it sees them) to do so. This acts as a constraint on the ability of the government to control BNOC and as a constraint on the ability of the government to maintain flexibility and adaptability. This is one of the areas in which the theory of bureaucracy can make a contribution to the understanding of government involvement in the North Sea.

The debate over the establishment of BNOC suggests that BNOC was not meant to be a means to an end but more likely, an end in itself. As Rhenman and Normann argue,[4] policy problems today are concerned with the choice of a strategy that can adapt to a continually changing environment. It has been largely because of this that planning in the North Sea has been institutional and pragmatic – which itself has created uncertainty. This chapter is not directly concerned with whether BNOC is an 'effective' organisation, rather the concern is whether any such organisation is able, within the democratic framework, to provide a positive contribution to the government and/or to the industry in a way that is not overwhelmingly political. BNOC's reactions to changing circumstances are analysed and the political nature of BNOC's strategy is considered.

5.2 THE POLICY DEBATE

The 1973–4 international oil 'crisis' heralded a new phase in the UK government's offshore oil policy. Apart from the balance of payments effects of the quadrupling of the price of oil, the most serious potential danger to the UK, and to many other industrial countries, was perceived to be the loss of secure supplies of oil. Even with the government's shareholding in BP it could not persuade BP to favour the UK in its allocation of oil. This experience raised serious questions with respect to North Sea oil as the licensing terms did not give the government significant control over either the rate of extraction or the destination of the oil once it had been landed in the UK. It was in this atmosphere of world crisis that considerable pressure was exerted on the government to instigate fundamental changes in offshore oil policy. Energy and especially North Sea oil, had become an important political issue and political parties were keen to be seen to be active in this area. The Labour Government's 1974 White Paper[5] was produced, significantly, only after the North Sea had become a major political issue. The preceding decade was characterised by the relative freedom from government intervention of the oil industry. The 1973–4 oil 'crisis', the discovery of sizeable oilfields in the North Sea in the early 1970s and the publication of the Public

Accounts Committee Report (Chapter 7) had the effect of pushing North Sea oil to the centre of the political stage. Individuals became 'aware' of the economic potential of North Sea oil. As the cost of acquiring information to the individual diminished (because the debate was taken up by the media) individuals' desire for information increased (when the importance of North Sea oil became increasingly apparent); their coercion threshold was lowered (see Chapter 3). The combination of these factors meant that it was in the interests of the political parties to formulate an oil policy which would attract the most political support. As North Sea oil became increasingly sensitive, politically, there was an increase in the potential political gains of being seen to be taking action. The dependence on OPEC oil and the 'excessive' profits being made by oil multi-nationals (MNCs) were prime areas where government action could result in political gain.

As oil prices rose the existing tax and royalty arrangements were not considered adequate and it was estimated that unless the conditions were changed the government would never take much more than about one half of oil company profits and about one half of the post-tax profits would be remitted overseas[6] (see Chapter 7).

It was therefore in this political climate that the Labour Government published the 1974 White Paper and the 'energy debate' which was given popular political support, began in earnest. According to Downsian theory[7] politicians are generally unaware of individuals' policy preferences due largely to the lack of incentive to the individual to become informed and to make that informed preference or opinion known. As previously discussed (Chapter 3), Breton[8] maintains that it is the very nature of the democratic system that 'shields' politicians from citizens' preferences (i.e. due to logrolling, information costs, the election period, etc.). However, once there is a perceived need for, in this case, a domestic oil policy, then in the formulation of that policy other factors (i.e. bureaucratic interests, intraparty rivalry, pressure groups, inter-departmental rivalry, etc.) become important. It was at this second stage that the government's principal objectives and considerations were outlined in the 1974 White Paper (see Chapter 2).

The proposals in the 1974 White Paper[9] were significantly different from the policies finally implemented by the Labour Government. This discrepancy is due, in part, to political and bureaucratic pressures which are an intrinsic part of the democratic process in the UK. By outlining these pressures it is hoped that a greater understanding of the limitations of the government policy process will result. The distorting effects within the policy process are not necessarily short-comings of the system. It may be that these pressures act as safe-guards restraining the 'enthusiasm' or 'excesses' of governments, (Chapter 8 argues that this is a mistaken argument). However, the point is that a discrepancy does exist between policy intentions and the final policy and secondly, between the intended results of policy and the actual results – and to a certain extent these discrepancies are inevitable.

Labour policy pronouncements and proposals concerning the establishment and rôle of BNOC made in 1974 and 1975 can be seen as part of a bargaining process between government and the oil companies. Initial Labour Government intentions for 51 per cent participation (see Chapter 2) can be seen in terms of strategic behaviour within the policy process. It is unlikely that 51 per cent participation was ever considered by the government as a rigid requirement.

This view is supported by the fact that even before BNOC was officially established on 1 January 1976, Lord Kearton (the first Chairman of BNOC) was referring publicly not to 'participation' but to the less emotive concept of 'partnership'.[10] This is consistent with Riker and Tullock's[11] idea of intra-party rivalry as a constraint on government output. If the straightforward 'socialist' action of 51 per cent participation as was originally proposed was implemented by the government there would not only be considerable opposition from the oil industry but also political opposition in Parliament and in the country. Although the government would remain in office it would run the risk of adverse and embarrassing political publicity and it would also allow the Opposition the chance to adopt a more popular policy (i.e. the constraint of political competition). As the government is not quite sure of voter preferences anyway, it is keen to minimise risk and take the safest option, while at the same time placating its own left wing by political rhetoric. The 'safest option' in a two party system would thus be one which tends towards the position of the other major party, converging on the centre. The constraint of intra-party rivalry (of Tullock and Riker) is more convincing than Downs' party competition argument but they are not entirely independent of each other. Given a government with a safe majority and an election period of five years, once in office the government will for a time be more concerned with revolts from within its party than with Opposition attacks. However, internal disunity may provide the ammunition for Opposition attacks and the cumulative effects of this over a period of years may add to the likelihood of defeat at a General Election. Of course if the government relies on the support of a minority group in the House of Commons then party competition becomes of crucial importance.

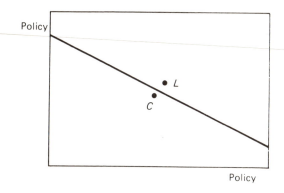

FIGURE 5.1   *Vote maximisation and inter-party competition*
SOURCE   G Tullock The Vote Motive (London: IEA, 1976) p. 21.

In Tullock's example, Figure 5.1, it is suggested that in a two party system the two parties (*L* and *C*) would tend towards each other near the centre of the distribution of voters, thus approximately splitting the voters 50:50. For the purpose of this paper, although the result is interesting, the crucial implication is that the politician (or in this case the political party) is attempting to adopt a policy in order that his chance of re-election will be maximised and not a policy that is likely to maximise the 'national interest'.[12] Thus, in establishing a national oil company, the Labour Government was careful on the one hand not to move too far away from the centre of the issue space diagram and risk party competition and on the other hand was careful to consider its own left wing so as not to risk a breakaway group (see Chapter 3).

As stated in Chapter 2, there was a further government ambition associated with the establishment of BNOC. John Smith, the Parliamentary Under-Secretary of State for Energy, argued that it was vital for the government to gain expertise and knowledge of all the everyday aspects of the offshore extraction industry.[13] Both the Labour Government and the Conservative Opposition were of the opinion that the government's ability to handle oilfield information and the quality of that information should be improved. Divergence of opinion between the two main political parties centred around the method by which this could be achieved. On the one hand the Conservative Opposition felt that a NOC was unnecessary and all that was required was a small regulatory body with power to gather information and to hire technical advisors to process that information. On the other hand, Labour argued that the gathering of information would be only one function of the NOC.[14] Thus the UKOCA proposition[15] was not adopted. When Mr Benn succeeded Mr Varley as Secretary of State for Energy he emphasised the need for first-hand information concerning North Sea activities and that this could not come about without large-scale direct participation. Mr Benn's argument was that because North Sea oil was of such crucial economic and political importance to the nation it could not be left to the vagaries of market forces nor could the government be dependent on others for its information and expertise[16] This argument is often used as a justification for government intervention in 'strategic' sectors of the economy. The Labour Government saw its rôle in the North Sea as one of 'controlling' rather than 'regulating' the industry. It is thus necessary for the controlling system to be able to steer the controlled system using various instruments and techniques. It follows that the controlling body must have more information and expertise than the system to be controlled, otherwise there is a danger that the rôles will be reversed so that it is BNOC which steers government policy.[17]

Thus the need for control over North Sea oil development and disposal was a principal objective accepted by both major parties. However, the method by which this was to be achieved differed greatly between them. Extensive powers were given to the Secretary of State for Energy in the Petroleum and Submarine Pipelines Act to regulate production in the 'national interest', the only constraints being the guidelines outlined by Eric Varley[18] by which the implementation of this authority would be governed, setting limits on the timing and extent of production restrictions (see Chapter 2).

A significant characteristic of the 1974 White Paper which tends to recur in later government policy documents is the extensive powers given to the Energy Secretary in order that the Minister may safeguard the 'national interest'. This authority granted to the Minister and hence to his senior advisers in the Department of Energy is essentially concerned with the personal power and political standing of the Minister.

Niskanen[19] implies that politicians and government bureaucrats attempt to solve 'problems' by redefining them and favouring the expansion of policies and powers to deal with them. The 'national interest' objective is open to various interpretations and can include a variety of goals; at the same time it is a politically appealing concept. Increased government control gives more power to the government bureaucrat and more prestige to the Minister in that Department. The licensing decision laid down the foundations upon which the government could intervene in the oil industry when it saw fit. In the process of adopting a policy Niskanen[20] argues that bureaucrats will favour economic planning because the influence – and size – of the department will increase as will its budget and all the personal benefits this implies. Much of the justification for BNOC does not stand up to the criticism levelled at it unless it is accepted that the 'national interest' criterion can be achieved by government intervention. The two Downsian assumptions with regard to the motivation of politicians (i.e. that they seek to maximise votes and they formulate policies by which to hold office) suggest that politicians will not attempt to impose policies that are in the 'national interest' unless the policy is also in direct accordance with the politician's self-interest. The Labour Government's view in 1975 that state intervention and control in the North Sea was by definition in the national interest because it was a socialist move does not explain the final characteristics of the policy. The original proposal of a policy may have been concerned with party ideology but the practical institutional problems are such that in order to implement a policy compromises with the bureaucracy and the industry have to be made.

One of the functions of BNOC, in common with many other European NOC's, is to forward the 'good' of the whole country,[21] not just the oil sector, for instance, by creating employment in politically sensitive areas, increasing government revenue and strengthening the pound. The inescapable consequence of such a wide framework of reference is that BNOC will be prevented from taking decisions on purely financial and commercial criteria. It is hard to justify BNOC as a sole licensee and operator not acting commercially, 'for example, for it [BNOC] to explore and locate reserves and yet not develop them'[22] and at the same time acting in the national interest.

Objectives with respect to pipelines, safety, good oilfield practice and the need for information and regulation, were acceptable to the Opposition. However the Conservative Opposition and the oil lobby (represented collectively by the UK Offshore Operators Association and individually by company executives) for different reasons, argued that the bureaucracy involved, the expense, the extensive personal powers of the Energy Secretary and the extent of BNOC's rôle in the

North Sea (especially with regard to participation) were unnecessary. Patrick Jenkin, then Shadow Energy Secretary, criticised BNOC for attempting to be both 'like an investment holding company – passive, watchful, steering, monitoring and supervising' and an 'active, entrepreneurial body which would aim to compete fairly with existing companies'.[23] This is a serious and genuine criticism of the rôle of BNOC in that, from the outset, there seemed to be a conflict between its activities as a competitive, commercial oil company and as a partner of the private companies gathering confidential information and with a specially intimate relationship with the government. BNOC was created to deal very closely with the government and although its precise relationship with the government was difficult to delineate there was considerable oil company opposition to the special privileges given to BNOC.

This opposition had an effect most noticeably with respect to participation negotiations, but also in other areas such as the loosely-adhered-to rule of landing oil and the refining of up to two-thirds of North Sea oil in the UK. These rules gave the Government the legislative powers for intervention and control to occur in the future.

The oil companies had a significant rôle to play, as pressure groups, in determining government policy with regard to the establishment and functions of BNOC. Downs[24] maintains that governments tend to favour the interests of producers over those of consumers. This is due to political parties being sensitive to any well-informed, well-articulated opinion that is voicing a preference. Producer pressure groups generally have strong common objectives, good organisation and are able to approach senior politicians and bureaucrats. Consumers' interests are much more diverse and thus generally less organised and effective as pressure groups. Breton[25] adds to the strength of this argument by suggesting that producer organisations (e.g. UKOOA) have an intrinsic need to survive which necessitates their seeking causes to champion, and being seen to do so by their members (see Chapter 3).

5.3 THE PETROLEUM AND SUBMARINE PIPELINES ACT 1975

The Petroleum and Submarine Pipelines Act, which became law on 12th November 1975, together with the 1975 Oil Taxation Act, implemented the powers and policies proposed in the 1974 White Paper; but was subjected to the effects of bureaucratic pressures, lobby groups, intra-party and inter-departmental rivalry in its passage through Parliament. The Act did not give BNOC a direct regulatory rôle (which Lord Kearton maintained it never wanted, and the Department of Energy was keen to keep)which was to be solely the job of the Department of Energy, which, with the Secretary of State for Energy, would keep in very close contact with BNOC. The Corporation would have less autonomy than most other nationalised enterprises. The close links between BNOC and the Department of Energy and the fact that the Department's influence was not diminished (or transferred) by the existence of BNOC meant that the Department retained its regulatory functions and did not actively oppose the establishment of BNOC. BNOC was empowered to search for and exploit petroleum resources anywhere in the world, to take over the government's participation interests, to buy and sell petroleum and trade in its derivatives, and to build, hire or operate refineries, pipelines and tankers.[26] BNOC would immediately take over the NCB's petroleum interests, thus bringing it into partnership with oil MNCs such as Gulf and Conoco, and this also gave BNOC immediate stakes in the Thistle, Dunlin, Hutton and Statfjord oil fields. The financial structuring of BNOC was highly controversial and was affected by inter-department rivalry largely between the Treasury and the Department of Energy as to where the control should be. This is not an uncommon problem with regard to nationalised enterprises in general.[27] BNOC could borrow money from the Government or with the consent of the Energy Secretary, from any other bodies at home or abroad. This amount should not, 'in normal circumstances', exceed £600m and the Energy Secretary was only able to raise it to £900m;[28] BNOC's profits would be paid into the newly formed National Oil Account (NOA) along with royalty proceeds. Thus together with its exemption from Petroleum Revenue Tax (PRT), BNOC would have an investment capability of over £2b, made up from £900m from the NOA, £500–£600m from royalties and £500-£600m from its exemption from PRT.[29]

Thus the structuring of BNOC gave the Energy Secretary very great powers, not only with respect to the financing of BNOC via the NOA, but also with respect to depletion policy. The purpose of the 'Varley Guidelines' was to reassure the oil companies as to the extent that the government would exercise depletion controls. Nevertheless, the government's intentions were far from clear and the Energy

Secretary had virtually unlimited powers to control depletion rates, again in the vague interests of the national good. The oil MNCs, as a powerful pressure group, objected strongly to the personal powers of the Minister and attempted to limit his powers during the passage of the Bill. The original clause which gave the Minister power to specify maximum and minimum production rates was withdrawn but the Energy Secretary still had extensive powers in the area of production (see Chapter 2). Generally, few changes were made during the passage of the Act that did anything to placate the fears of the oil companies. Producers were instructed to submit programmes detailing their capital investment plans and maximum and minimum annual production rates for oil and gas.

The Energy Secretary has the power to reject programmes if they were not in the 'national interest' or were considered not to be consistent with 'good oilfield practice'. Both these criteria can be interpreted so as to encompass almost any conceivable circumstances that the Minister may include. Additionally, even if the programme is approved by the Energy Secretary, he can give the producer a 'limitation notice' which specifies limits within which the Secretary can, by using a further notice issued after some designated period, direct him to produce. Another notice may require either a speeding up of the producer's depletion plans, in the case of a 'national emergency', or a slowing down in order to protect the 'national interest'.[30]

Added to the Minister's crucial powers on production the Minister had control of all aspects of North Sea submarine pipelines. This had great potential importance as companies would have to ask for permission to use existing pipelines, thus giving the government leverage over companies in order to persuade them to follow government policy.

Further powers to the Minister accrued due to BNOC's exemption from PRT. The Government looked upon this as a legitimate method of providing finance for BNOC as the money would simply be directed to a different government department and the purpose of PRT anyway was to ensure the nation was receiving a 'fair share' of North Sea profits. Thus BNOC's exemption from PRT was, according to the government, 'an internal administrative measure to simplify transactions within the government'.[31] Criticism from the Opposition was predictable: it would create a distortion of investment criteria between two partners, in a single field, with a single licence, who have to arrive at a common decision in selling oil downstream. In competition with other companies BNOC could be tempted to shade its prices in a way that would be unfair to its competitors. Also, exemption from PRT could be used as a way of concealing the size of the borrowing requirement needed to finance BNOC. Although the annual accounts of BNOC would be drawn up to show what would have been the effect of paying PRT, the main criticism was that it would in effect be a bogus profit by-passing normal parliamentary control exercised via the Treasury and it would give more power to the Minister.[32]

Exemption from PRT compounded another fear of the oil industry; it would make financial comparisons between BNOC and other companies more difficult. This was a further area where the imprecise, vague rôle of BNOC resulted in confusion due to its conflicting activities as a competitor and partner in the North Sea. By giving BNOC general and non-commercial functions to fulfil, gauging its success on purely financial or commercial criteria or with comparisons with other oil MNC's is inadequate. Thus one of the political 'benefits' of NOC's wide framework of reference is that judging their performance becomes very difficult.

The wide-ranging personal powers bestowed on the Energy Secretary by the Petroleum and Submarine Pipelines Act are generally consistent with the techniques (outlined by Niskanen)[33] employed by bureaucrats and in this case politicians, in increasing their influence and power. At this stage there is only limited conflict between bureaucrats and politicians because their ambitions are broadly similar. The passage of the Act was relatively trouble-free because it was in line with Labour Party ideology, there was strong popular support for the government to be seen to be taking some sort of action in the energy sector and the Act increased the stature of the Department of Energy. The combination of these factors meant that Conservative and oil company objections were relatively ineffectual. Consequently a Conservative Government could be expected to have greater difficulty in formulating an alternative policy (or changing an existing one) because reducing the influence of Central Government – which is an important aspect of Conservative Party ideology – could, but not necessarily would, contradict the ambitions of an individual minister and, more likely, would meet with opposition from the Department of Energy whose activities would be reduced or restrained. These conflicts would arise before considering the opposition from the newly formed bureaucracy of BNOC and conflicts with other macro-political priorities. To an extent, as will be discussed below, the Conservatives have encountered these problems since 1979.

In BNOC's case, there was still accountability exercised through the NOA but there was greater discretionary power in the hands of the Minister than is usual for nationalised enterprises. The Minister, at the end of each financial year, had to report to the Comptroller and Auditor General (CAG) who in turn reported to the Public Accounts Committee (PAC) of the House of Commons. Concern over BNOC's use of royalty money was voiced in 1977 by the CAG as royalty monies became available to BNOC without parliamentary control. If BNOC was to rely on the royalty monies instead of on outside borrowing, then BNOC would be subject to control not by Parliament but by the Energy Secretary in its use of these funds. If BNOC had not been given privileged access to the royalty monies in the NOA it would have had to borrow from the National Loans Fund (NLF) like any other public corporation, in a manner subject to conventional parliamentary control. Also, this sizeable source of financing was technically a forward sale of oil and not a borrowing, thus it did not count against the borrowing limits monitored by Parliament. A further shortcoming of BNOC's financing-accountability system was that although the CAG had direct access to the NOA he only had indirect access to BNOC's internal accounts and records.

Following the PAC's 1977 report BNOC announced that their funds would be split 60:40 between equity and non-equity monies. The royalty cash from the NOA, to which BNOC's retained profits would be added, would be equity funds until they reached 60 per cent of the total, after which they would be divided between equity and non-equity funds so as to keep to ratio 60:40, taking into account any other long-term borrowings or outside non-equity finance. The result was that BNOC was put on a financial footing broadly similar to that of a private sector oil company.

5.4 BRITISH NATIONAL OIL CORPORATION 1976–1979

With the formal establishment of BNOC on 1 January 1976 the third stage of the government policy-making process is reached. The factors influencing the final outcome of the policy are not exclusive to this stage and the actual beginning of this stage would have been sometime in 1975. A prime consideration with respect to the establishment of BNOC was the need to maintain flexibility in order to be able to adapt and adjust to a changing economic and political environment. The Petroleum and Submarine Pipelines Act laid down the legal and institutional foundations upon which the government's oil policy would be implemented, leaving adequate scope within which the government could respond to unforeseen developments.[34] One of the consequences of this was that BNOC did not have explicit long-term objectives.

A characteristic of this stage of the policy process that differentiates it from the formulation of policy stage is that with the existence of BNOC there is an added set of relationships and interests to consider. The relationships between BNOC and the oil MNCs and between BNOC and the government had the effect of altering the structure and nature of the offshore oil industry. BNOC employees now had their ambitions and objectives with respect to BNOC's future which were distinct from ambitions and objectives of government bureaucrats and politicians who created BNOC.

In its first report[35] BNOC outlined its wider aims and saw its position as combining the functions of an instrument of national policy, of a commercial enterprise and of an adviser to the government. Within this ambitious framework BNOC sought to become an efficient manager of its equity interests in offshore petroleum exploration, development and production. It also aimed for the effective disposition of the petroleum available to it from equity interests and from participation arrangements. Moreover, BNOC intended to develop its expertise particularly in the area of the development and use of resources on the UKCS and to be able to advise the Secretary of State on all oil matters. BNOC's immediate objectives in late 1975–6 were, first, to obtain a competent staff and second, to work out participation arrangements with the oil companies.

As has been explained (see Chapter 2) the government's original intentions with respect to participation had been abandoned in the face of oil company opposition. Lord Kearton had to construct a method of participation which the oil MNCs could tolerate and which still conveyed enough of the 'spirit' of participation to satisfy Labour politicians concerned with socialist ideology. Thus what subsequently remained of participation was based around the need for government bureaucrats in the Department of Energy to acquire information and also for a say in how the oil companies disposed of the oil they produced.

Participation was redefined to mean that BNOC was to be treated as an equitable partner in that BNOC would have access to information but would have no equity shares. 51 per cent oil was conceded to BNOC on the understanding that BNOC would hand back to the companies all financial and other benefits (the oil would be purchased by BNOC at the market place). BNOC would be a member of the operating committees that manage fields but 'there was no question of exercising a 51 per cent vote'.[36] Lord Kearton described the arrangement as BNOC having a 'meaningful' vote which he defined as being a rôle where BNOC's voice would carry weight but BNOC would have no power to direct the pattern of development or veto decisions. However, under the Petroleum and Submarine Pipelines Act the government still had the power to review and reject development programmes.

Lord Kearton supported this somewhat vague arrangement as the principle around which individual participation negotiations could take place. The early controversies illustrated Lord Kearton's practical and pragmatic approach to his job of establishing BNOC as a serious and powerful NOC. He stressed that BNOC would have access to, and control over, 51 per cent of the oil and it would be able to argue its case from the inside, thus having more influence and giving a greater and faster contribution than it would have done from the 'outside' with little knowledge or information. The 'new' definition of participation from which individual agreements could be worked out would give BNOC an opportunity to gain acceptance and credibility in the eyes of the North Sea oil companies.[37] This was of crucial importance to BNOC when seen in the context of its longer term ambitions. In these early stages of BNOC's existence its immediate ambitions and those of the government were in close accordance and this accounted for much of the early success of BNOC. Evidence[38] from other European NOCs has shown that unless an NOC has the backing of a committed government in its early years the NOC will tend to lack direction. The Labour government backed BNOC enthusiastically and Lord Kearton had a strong personality with both experience in dealing with politicians and a proven business ability. Although BNOC met opposition from the oil MNCs with respect to participation the combined efforts of the government and BNOC were able, eventually, to reach agreements with all the North Sea companies.

Whether or not BNOC would fulfil the prediction that it would develop into a self-perpetuating bureaucracy which would attempt to 'steer' government policy rather than be steered by it was not apparent at this stage. The activities of BNOC were uncontroversial in that they were in line with government expectations. However, by looking at BNOC strategy over the first five years of its existence there appear to be noticeable trends consistent with the hypothesis of BNOC seeking independence from government control and acting as an oil MNC. A participation deal agreed with BP[39] illustrated Lord Kearton's determination to establish BNOC as a fully integrated oil company with avenues open to it, if it should choose, to expand downstream. The agreed formula with BP included provisions for collaboration on downstream activities, and also for the provision of training facilities for BNOC staff. Whilst this was in line with the government's objectives for BNOC it was also consistent with BNOC's longer term strategy to become independent.

A genuine source of apprehension on behalf of the oil companies was with respect to the precise rôle of BNOC which was still somewhat vague. Lord Kearton's job in establishing BNOC was conducted in a cautious manner as he had to encourage the oil MNCs into participation agreements, whilst at the same time future development could not take place at the desired speed if the (US) MNCs were 'forced' to leave the North Sea; thus no two participation deals were the same. The oil MNCs thus still had considerable bargaining power and were prepared to oppose the implementation of the stated government policy as long as possible. It is significant that the first participation deals with BNOC were all with relatively small companies which were generally in some sort of financial difficulty or needed government assistance in another way (i.e. deals involving United Canso, Deminex, Champlin Petroleum, Berry Wiggins, Consolidated Gas Fields and Tricentrol).[40]

Within the framework of BNOC's government-imposed long-term objectives there was some confusion as to the immediate rôle of the Corporation and how it would reconcile two apparently conflicting functions. As a partner with the oil companies it would have access to restricted, confidential information and at the same time would be attempting to act as a competitive rival in the market. Over time, of these two functions, it became apparent that whilst the government supported this duality of rôles, BNOC – once it had become firmly established – was overwhelmingly concerned with the latter objective. It is in this area that Niskanen's theory of bureaucracies can be applied to the relationship between the government and BNOC; this will be examined below, following a description of BNOC's activities.

The Petroleum and Submarine Pipelines Act had given BNOC the legal and institutional potential to expand into virtually any sector of the international oil industry. The direction its activities were to expand into were subject to ministerial ratification but the onus was largely on Lord Kearton to concentrate on such areas as he saw fit. In 1976 his 'overwhelming priority' was to gain access to more oil, that is, to increase BNOC's equity share in the North Sea to between 30 and 40 per cent.[41] BNOC's desire to move through the 'learning curve' as rapidly as possible, in order that it might act as an equal with its private company partners, was seen as justifying its presence on operating committees, thus allowing BNOC personnel to build up their expertise in their specific sphere of operation and in the oil development business in general. The acquisition of the NCB's and Burmah's North Sea subsidiaries was another way by which BNOC gained experienced staff. Thus these activities, whilst in direct accordance with government ambitions for BNOC, were also enabling BNOC to fulfil its own objectives.

Lord Kearton was careful to point out that BNOC was in no way acting as a regulatory agency and the Department of Energy alone, for example, decided to whom to grant licences and upon what criteria their decisions would be based. However, Lord Kearton made it clear that he felt the government was right to come to BNOC for advice and this naturally seemed to the private companies to put BNOC in a privileged position. Lord Kearton admitted that in some cases the Department of Energy had asked BNOC's advice on the variation in a licence[42] and whether or not they would be interested in it themselves, and BNOC had accepted their offer. Thus to an extent the oil companies' fears were justified.

A further example of BNOC's privileged position arose out of the terms of the Fifth Licensing Round in the form of a disagreement between BNOC and UKOOA.[43] Whereas North Sea operators had to wait for allocation rounds, BNOC had the right to apply for any unlicensed block at any time. New BNOC partner companies saw this as putting them at a serious advantage. If an oil or gas reservoir was discovered in one of the new UK blocks and it was thought likely to stretch into an adjacent but unlicensed block, only BNOC could bid for that block immediately. Thus the partner companies who helped finance the find would lose out. Thus even at the implementation stage of the policy oil company pressure groups were still attempting to change the policy or at least delay its implementation until a possible change of government. Their success was limited (although they did achieve various concessions in the participation negotiations) largely because the decisions to give BNOC extensive powers and to increase the ability of the Department of Energy to control North Sea activities had already been taken and had the backing of the government, the Department of Energy, BNOC itself and apparently the voter (since they seemed to fit prevailing public opinion).

Details of specific policy points were still unclear but since the government, BNOC and the Department of Energy all had the same short-term objectives, the oil MNCs (with a sympathetic Conservative Opposition) could do little other than adopt a position of non-co-operation and delay.

These tactics of non-co-operation were employed by the oil companies most noticeably in participation negotiations. Lord Kearton[44] conceded that his team of negotiators sometimes had to wait months for an oil company reaction to a government proposal and although this would have, in part, been due to oil company negotiators having to report in detail to their superiors, and oil companies operating in consortia having conflicting interests, it seems that these delays were symptomatic of the oil companies' general opposition to BNOC.

Negotiations with US oil MNCs were particularly sensitive because if terms were too unfavourable these majors could decide to leave the North Sea, creating a vacuum which BNOC was not yet able to fill. The government would also have to consider possible retaliatory measures taken against UK companies (specifically BP) operating in the US. The government could decide to exclude oil companies from future licensing rounds if the Department of Energy felt the oil companies were being too unreasonable – again discretionary power on behalf of the Ministry was considerable. Amoco's refusal to enter a 'voluntary' participation agreement resulted in their being totally excluded from the Fifth Round allocation of licences.[45] A further consideration, which is much harder to substantiate, is the possibility that oil MNCs which created 'unacceptable' obstacles to reaching agreements on participation could suffer a more general loss of government sympathy. This loss of 'goodwill' could manifest itself in any future negotiations with the government on oil-related areas where the Minister exercised arbitrary powers or rules were open to interpretation. These companies could, for instance, be descriminated against in the allocation of blocks in a licensing round, but this is impossible to substantiate because of the nature of the allocation procedure.

In the fifth licensing round BNOC was given a 51 per cent share in all 44 blocks allocated and was to be the operator on four of them. In

August 1976 71 blocks, or part blocks, were offered and in February 1977 24 offers of licences were made. In the four blocks where BNOC was the operator it would be operating on behalf of Shell and Esso, BP, Hamilton Brothers and Kerr McGee. BNOC had to bear 51 per cent of the exploration costs on approximately 9000 square kilometres of sea bed and before any licence was issued the prospective licensees had to conclude with BNOC an operating agreement acceptable to the Secretary of State. These activities necessitated BNOC seeking considerable finance, which it obtained from commercial US and UK banks.

This was a highly significant episode in the development of the Corporation. BNOC set up an American corporation called Britoil which borrowed the cash ($825m in all) to pay in advance for quantities of North Sea oil pledged by BNOC from the third quarter of 1978 until mid-1981. BNOC was obliged to deliver enough oil to pay the interest on the loan and between 1981 and 1985 BNOC would deliver oil in sufficient quantities to pay off the principal as well. This deal[46] enabled BNOC to obtain finance with no government guarantees and BNOC claimed the terms it was given as a borrower were as favourable as any international oil company, illustrating the confidence the financial institutions had in BNOC. Of the total, $675m was at rates fractionally above Citibank's prime rate; the rest was 1 per cent above the London inter-Bank rate, giving an average then of 7½ per cent. At the end of 1977, BNOC's financing costs were equivalent to an interest rate of less than 8 per cent representing a saving of over 6 per cent compared with the borrowing from the NLF. At the time of finalising the deal (in mid-1977) BNOC was borrowing just under £350m from the NLF at around 14½ per cent. The new dollar loan allowed BNOC to pay off these government loans and reduce its immediate interest costs.

5.5 THE APPLICATION OF NISKANEN'S THEORY OF BUREAUCRACY TO THE DEVELOPMENT OF BNOC

In examining BNOC's performance in the context of Niskanen's theory of bureaucracies, there are basic differences which have to be made clear. BNOC's relationship with the government is a relationship with both a government department and with government politicians. It has previously been stated that the nature of the relationship between a government department (the bureaucracy) and the politicians (the sponsor) will lead to distortions due to conflicts of objectives. It would thus seem to contradict this if it was now assumed that BNOC is the 'bureaucracy' and the government, comprising both politicians and civil servants, is the single sponsor. To begin with it is sufficient that the government department and the government can be assumed to be, for simplicity, one sponsoring body. This assumption will be relaxed subsequently. This problem again emphasises a significant consequence of the creation of BNOC; an additional institution is established within the policy process with its own set of objectives and characteristics.

As has been stated (Chapter 3) the power of a government bureaucracy stems from the nature of its relationship with its sponsor, the government. The creation of BNOC was a major political event and BNOC was explicitly designed to have an especially close relationship with the government, more so than other public enterprises. Although individual civil servants may spend a considerable amount of time dealing with the affairs of BNOC, the higher level civil servants (like politicians) have numerous functions to perform and have many other areas that need their attention. Employees of BNOC, however, are working full time in the interest of themselves and BNOC. This argument is not as forceful as in Niskanen's case of the politician being only a part-time overseer of a department's activities. In 1977 the Department of Energy was reported[47] to have forty-one officers of Principal level and above involved in North Sea oil and BNOC matters, of which eighteen were experienced in North Sea oil business. Nevertheless, the relationship can still be viewed as similar to Niskanen's – one of a bilateral monopoly with BNOC having access to considerably more expertise and information than does the government. If the government is seen not as a single entity but as a government bureaucracy and a government politician, then although the government bureaucrat will be more able to acquire information and have a stronger relative incentive to do so than the politician, BNOC is nevertheless in a very strong position with regard to the amount of, and standard of, information it passes on to the government's department. Even then however, the department may attempt to withhold information from the politician if it is in its interests to do so.

The financial structuring of BNOC gave it the opportunity to acquire funds from the Energy Secretary or from commercial banks. Obviously, to begin with BNOC was dependent on the government for cash but the government was keen to support BNOC and was

sympathetic to its demands. Moreover, given that BNOC was re-
cruiting high-skilled labour it was in a position to make a strong case
for itself to the government. It is important that negotiating would
have been carried out in a favourable atmosphere as the government
was keen for BNOC to establish itself quickly alongside the oil
MNCs.

Thus the relationship between BNOC and the government would
be seen in terms of a Niskanen-type sponsor/bureaucracy relationship, but with significant differences. The most important difference
is that in this case the sponsor cannot be seen as a homogeneous
entity with one set of characteristics and objectives based on a single
rationale. Of less importance is the fact that BNOC was able to seek
finance from any source. BNOC could only negotiate for commercial
loans once it had proved its credibility. This initially it could not do
without government support and thus BNOC was totally dependent
on the government for funds in the first two years of its existence.

Criticism by Breton and Wintrobe[48] of Niskanen's model is also
evidence here. The government could not in BNOC's case be viewed
as a passive sponsor because accountability was exercised through the
NOA. Two points are significant here; firstly, that the Energy Secretary had extensive personal powers over BNOC's finances and
secondly, without privileged access to royalty monies in the NOA,
BNOC would have had to borrow from the NLF and have been
subject to conventional parliamentary control. If there had been no
NOA, the government would be able to acquire some knowledge of
the operations of BNOC via the PAC and also through the Department of Energy and this would act as a constraint on BNOC.
Breton[49] has argued that the existence of controls must mean that
they work at least partially. This is not necessarily true as the greatest
need of a politician is to be seen to be controlling the excesses of, in
this case, BNOC and political 'safety' can be acquired by the existence of controls and not by their effective use. To gauge the efficiency
of BNOC the CAG would have needed very detailed information on
all aspects of BNOC's activities and would have had to set precise
objectives to be able to judge BNOC's success. Because of the wide
framework of reference imposed upon BNOC, financial and commercial comparisons became almost impossible. A profitable public
corporation is a valuable political asset, especially for a Labour
Government, and if BNOC was seen to be making a sizeable contribution to government revenue it might have tended not to be scru-
tinised as closely as if it had been losing money.

Niskanen assumed the bureaucrat to be a rational utilitarian who, because of the nature of the sponsor/bureaucracy relationship, rationally attempts to maximise the size of his budget. The government/BNOC relationship had similar characteristics of a well-informed bureaucracy and a 'semi-passive' sponsor. However, assuming a rational utilitarian manager in BNOC, his maximand would not have been to maximise the size of the budget received from the government because other factors in the manager's utility function would not necessarily have increased as BNOC's budget increased; or more specifically, other factors would probably have had a higher ranking in his utility function. Increasing its size and growth rate were prime concerns of BNOC but it was how this was to be achieved (and the reasons why) and the implications of alternative methods of financing that were important.

The political rôle of BNOC has been emphasised in this chapter. BNOC was a product of Labour Party doctrine and its structure reflected the Labour Party's belief in the need for the central government to influence the economy. From the very inception of BNOC the Conservative Opposition made it clear they were strongly against the principle of an NOC and would dismantle the Corporation when they came to power. Thus BNOC executives in some form were overwhelmingly concerned with the survival of BNOC and the security of their jobs in the face of substantial political uncertainty. A second factor with regard to the motivation of BNOC executives is that, with the exception of political appointees at Board level, BNOC executives and managers would have had similar ambitions to those in private sector oil companies and would have had a natural aversion to the socio-political considerations imposed on them by Central Government. It is thus the assertion here that once BNOC had consolidated its position as a credible oil company the combination of these factors resulted in its being rational for BNOC executives to seek managerial independence which they were able to do, and were motivated to do, due to the nature of BNOC's relationship with the government.

Methods by which this ambition of managerial independence could be achieved were numerous.[50] One of BNOC's earliest priorities was to attract expert labour, skilled in the technical aspects of offshore oil. Skilled labour joining BNOC meant that BNOC's bargaining position with the government was strengthened and at the same time BNOC's standing in relation to the oil MNCs improved. A corollary of this point is that skilled labour may be less keen to join a NOC than they would an independent oil company. However, there seems little evidence to support this view and it is possible, as Lord Kearton maintained, that the reverse is true.

The most important technique by which BNOC could acquire managerial independence was by becoming financially independent of the government. The $825m borrowed from commercial banks gave BNOC a degree of independence from the government and at the same time was an added complication in any future Conservative effort to break up BNOC.

The power of Niskanen's model stems largely from the characteristics of the bureaucracy. Niskanen defines[51] a bureaucracy as being a non-profit organisation financed (at least partly) by a grant from its sponsor. The BNOC/government relationship differs fundamentally from the Niskanen model because BNOC was an income earning organisation and able either to finance itself through retained profits or by seeking a loan from an independent financial institution (i.e. not its sponsor, the government). Although BNOC needed ministerial ratification for many of its activities it was not operating under the same conditions, with the same restrictions as Niskanen's government bureaucracy.

The structure of the international oil industry is such that there is a trend for NOCs in general towards multinationalisation.[52] With respect to BNOC multinationalisation fulfilled two prime objectives: it decreased the opportunity for government intervention and it added to the complexities of dismantling BNOC. Similarly, long-term supply agreements and joint ventures – with foreign companies particularly – added to the difficulties of government control and there might also be technical difficulties with respect to auditing the accounts of foreign subsidiaries.

The greater the politicisation of BNOC and the closer its links were with the government the more it would fall in line with the Niskanen model. The 'national interest' objective could be employed to justify virtually any non-profitable project entered into by BNOC. Nevertheless, Niskanen's model still has serious implications for activities within BNOC itself and the desires of managers to increase their own stature by attempting to expand their departments or budgets or projects due to their relationship with their superiors. This would lead to inefficiency and the oversupply of output. For instance, because of the concentration of specific skills in investigating a project's usefulness, a team of experts could be able to persuade a superior of the considerable potential 'value' of a particular project:

i.e. they would be able to make out a convincing case for its being in the 'company's interest' to adopt the project. This is more than an 'unconscious' project enthusiasm; it is a conscious decision taken by a manager in order to maximise his own utility function which could be achieved by increasing the budget allocated to him or his department. This is not peculiar to BNOC but could exist in any large organisation. Management slack[53] is consistent with behavioural theories of the firm as well as Niskanen's closely related theory of bureaucracy.

A manager, not a top level Board member, will attempt to maximise a personal utility function consisting of, according to Williamson[54] salaries, staff, discretionary spending for investments and management slack absorbed as cost. The major constraint is that revenue generated must still be greater or equal to the minimum revenue demanded. Increases in salary are consistent with increases in staff but the manager's desire to increase the number of his direct subordinates will also be related to his own job security, prestige and managerial flexibility. Discretionary spending for investment reflects the ability of managers to direct resources to their own control, not on the basis of wholly commercial criteria.

The results of these behavioural factors are similar to the consequences of Niskanen's model (i.e. inefficiency and oversupply of output) but their sources are different and not specifically concerned with the bureaucrat's relationship with the sponsor and the bureaucrat's attempts to maximise his budget. Williamson's predictions of management slack were fulfilled, with respect to BNOC, because of its privileged position under the Labour Government. Here BNOC was, to an extent, protected from the forces of competition by the government and thus BNOC managers were able to satisfy their self-interest objectives without the constraints of competition. The major constraint was that BNOC had to be seen to be fulfilling political functions determined by the government.

Thus multinationalisation was a rational objective for the BNOC manager because it affected the survival of BNOC and the security of the manager's position. Multinationalism could also be supported as a strategy on commercial grounds; the economies of scale potential in the oil industry, the geography of oil supply and demand, and most importantly for BNOC the spreading of technical and political risk.

Relaxing the assumption of a homogeneous sponsor and thus considering the government as being made up of civil servants (bureaucrats) and politicians whose ambitions often contradict each other and cause conflicts and distortions, the oil executives' need for decision-making freedom becomes more apparent. Macro-political objectives of senior politicians might have conflicted with the long-term commercial aims of BNOC. If the government had been the sole financer of BNOC and BNOC had needed to request government money for an investment project the finance could have been denied because it conflicted with other government objectives and not because of its commercial viability. Because the Petroleum and Submarine Pipelines Act laid down the institutional potential for further government action and gave the Energy Secretary considerable discretionary powers, political uncertainty resulted.

Rivalry between government departments (most noticeably between the Department of Energy and the Treasury)[55, 56] can lead to government bureaucrats making decisions on the basis of whether they are losing some of their 'duties' to another department. BNOC, in seeking financial independence, was implicitly assuming that political interference (rather than regulation or even control) and the added uncertainties this created was likely to become much worse over time and at some stage would force BNOC out of existence or reduce its independence. Thus BNOC executives, on the basis of their rational expectations of the future were predicting political intervention and taking pre-emptive measures in order to protect their own interests. This not only caused difficulties for the decision-makers within BNOC but was an intrinsic shortcoming of government energy policy due to the characteristics of government and the policy-making process. It has been argued[57] that BNOC was a necessity because as the big oilfields of the UKCS were developed and only less attractive 'marginal' fields remained the oil MNCs would tend to leave the North Sea; thus a domestic NOC, in the national interest, had to be ready to fill the gap. An alternative view is that as the government makes the economic environment less attractive for the oil MNCs (e.g. by adding to the uncertainties through the creation of BNOC, by a punitive tax régime, by depletion controls) the government is actively forcing the oil companies to leave (thus justifying the existence of BNOC) whereas in other circumstances they would have remained.

5.6 THE 1979 CONSERVATIVE GOVERNMENT

At the beginning of 1979 the Conservatives were publicising their plans for the future of BNOC which would significantly have reduced

BNOC's influence in the domestic oil industry. This led to a great deal of speculation and uncertainty in the oil industry. The complexities of the financial structuring of BNOC and the substantial efforts made by Lord Kearton to make it as difficult as possible to break-up the Corporation, combined with a new 'appreciation' of the value to the incoming Administration that the political revenue of BNOC would bring, resulted in government procrastination and ambiguity with respect to BNOC.

By the beginning of 1979 supplies of crude oil on the world market were in turmoil with deliveries out of Iran at a virtual standstill. Despite OPEC production being higher in the first half of 1979 than in 1978 there was a perceived temporary excess demand for crude oil. Thus the volatile and sharply upward movement of oil prices during 1979 was at least in part due to uncertainty over supplies and the resulting changing expectations of oil prices. This uncertainty was accentuated by the decline in the volume of crude oil made available to the oil MNCs as sales by producer governments and NOCs took their place.

Thus at a time when there was great demand for low sulphur crude oil, BNOC, with no refineries of its own, had a healthy flow of easily accessible oil and with no restrictive long-term trading contracts BNOC was thrust into a position in the international oil market that was possibly out of proportion to its true standing. However, most of BNOC's oil was participation oil which it received on the understanding that the original owner would be left no worse off by the transaction than if it had sold the oil itself. In 1979 only about 15 per cent of BNOC's sales were of equity oil on which the Corporation takes all the profit. The volume of oil traded by BNOC increased from about 300 000 b/d in the fourth quarter of 1978 to around 800 000 b/d at the beginning of 1979 and over one million barrels per day by the end of 1979.[58]

These trading activities increased the status and prestige of BNOC and were factors in the revised view of BNOC's rôle taken by the Conservative Government. In 1979 the financial prospects of BNOC were good and together with its increased access to oil supplies the Corporation was able to put forward a strong case for its own survival. At a time when the US was subsidising oil imports, BNOC argued that it would be foolish to dispose of indigenous oil already under state control, simply to achieve a short-term boost to the budget. BNOC gave the government an alternative option in directing North Sea oil during a period of uncertain supplies. Also, and crucially, the cash generated by BNOC would be a very attractive long-term prospect to the Conservative Chancellor trying to reduce government spending.

The Conservative Government was finding it very difficult to reconcile important policy conflicts within its own election platform. The Conservatives came to power pledged to decrease state intervention, and specifically to dismantle BNOC. In the summer of 1979 the Energy Secretary (David Howell) was facing considerable political embarrassment in that a major oil producing country was suffering petrol shortages. Diverting resources to domestic refineries by the government would have resulted in a non-optimal allocation of resources and would harm Britain's position as a trading nation.

However, the government's prime consideration was the adverse and very public effect reducing oil exports would have on the balance of payments. Using BNOC as the instrument by which the government would attempt to divert oil supplies – whilst at the same time considering how to dismantle BNOC – would add to the embarrassment of a staunchly non-interventionist government. The choice to be made was not solely a political one.

The pressure to make use of BNOC for political purposes would come from within the Party and there would also be pressures from within the government to use BNOC revenue for top priority economic objectives (i.e. reducing the PSBR). These pressures could involve, for example, the Energy Minister having to be prepared to subordinate his own personal and departmental objectives to those of the government as a whole or to those of another Ministry. A Minister would not necessarily be averse to doing this if it was the case that by being a loyal government member his standing, and his prospects, were enhanced in the eyes of the party leader, party hierarchy or the party in general. This argument can be extended to include the actions of top-level civil servants. It could also be the case that government bureaucrats (at the most senior levels) and Ministers would be prepared to accept a budget reduction (the opposite of Niskanen's thesis) if it was in their personal interest to do so. Less senior bureaucrats would not necessarily have this opportunity to display loyalty and solidarity, and be rewarded for it, and would attempt to change the policy or prevent its effective passage. Thus although senior bureaucrats and Ministers may be able to increase their own standing by reducing the budget (and/or the size) of their department, for example by making the department a more tightly knit group of high-skilled experts, bureaucratic pressures from within would tend to oppose this.

The Conservative Government did consider it expedient to intervene in the oil industry despite its own proclamations to the contrary and the adverse economic consequences. BNOC, acting as the oil agent of the government[60] was instructed to persuade twenty small US and European refiners who were buying 75 per cent of BNOC's oil to export up to 50 per cent less. The Energy Department asked the major US and UK MNCs also to reduce their exports of North Sea oil.

Thus although committed to re-organise BNOC and to remove the special privileges bestowed on it by the Labour Government, the Conservatives were not slow to appreciate the usefulness of BNOC, to be employed on an *ad hoc* basis and also as an aid to fulfil medium-term economic aims. In July 1979 David Howell announced[61] that BNOC would be effectively split into two organisations and the slimmed-down corporation would compete on equal terms with the private sector oil companies. The two tier organisation would have its oil trading activities separated from its rôle as an explorer and producer. Mr Howell justified this decision by referring to the shortages experienced during the spring and summer of 1979 and, again, to the need to provide some assurance of secure supplies.[62]

The method of widening the ownership of BNOC was not settled but there were definite alternatives being considered by the government. All BNOC's shareholdings in producing fields or fields under development could be sold to other oil companies leaving a drastically diminished company which would have to rely on its exploration prospects for growth. Alternatively, all of BNOC's offshore assets could be retained and shares offered on the Stock Exchange. A major objective of the Conservative Government was the 'privatisation' of BNOC; the government hoped to sell off interests to the value of around £400m (out of BNOC's total assets of between £2 billion and £2½ billion).[63] The government found this very hard to carry out. As the Labour Government received opposition from various sources within the political system and also from external pressure groups and other interested parties with regard to the establishment of BNOC, so the Conservatives met with opposition with regard to BNOC's restructuring.

The task of dismantling BNOC was substantially more difficult than was the original task of creating BNOC. The need for a NOC in order to protect the nation from the uncertainties of OPEC and the (mainly US) oil MNCs is a politically appealing need. A certain amount of national pride is generated by a British NOC competing alongside the US giants, and (to an extent) dictating policy to them. A government subsequently in power that attempts completely to undo such a policy can be seen to be diminishing its own stature by resorting to 'petty party politics'. Confronted with such a problem the Conservative Government faced difficulties which are again intrinsic to the system. Reducing the central government's rôle in the economy by definition means that government bureaucrats will lose power and influence and many will be made redundant.

Thus there was considerable opposition from the government bureaucrats to a Conservative Government attempting to transfer a substantial part of the public sector into the private sector. In the case of BNOC, the bureaucrats' relative bargaining strength over the government was enhanced due to the political popularity of a NOC. Non-co-operation by civil servants can cause delays; a government Minister has numerous other important functions to fulfil and therefore must delegate responsibility to junior ministers and to senior career-orientated bureaucrats (who may be in opposition to the policy). By employing Niskanen and Breton-type techniques (see Chapter 3) of withholding and selecting information it would be possible to add considerably to the complexity of the task facing the Minister.

A further important source of opposition to the Conservative Government was from BNOC itself and the supporters of BNOC within the political system. Conservative opposition to the principle of BNOC had been well-publicised and their actions when they came to power were to a large extent, predictable. As has been discussed, the primary objective of BNOC between 1976 and 1979 (once having proved itself as a competent oil company by taking advantage of the Labour Government's enthusiasm for a domestic NOC) was to ensure its long-term survival by seeking managerial independence. Thus a definite strategy was adopted by BNOC in order to protect its own interests. Another tactic employed by BNOC was to make itself as useful as possible to the Conservatives – most significantly with regard to the revenue that would accrue to the government.

Thus the forces acting upon the government were such that it was unlikely that a Conservative Government would be able to bring about all the changes it had originally intended. The idea of completely dismantling BNOC was abandoned and various forms of restructuring and privatising BNOC were suggested. A significant point with respect to the privatisation of BNOC being achieved by a bond issue (as outlined by David Howell at the 1980 Conservative Party Conference), was the difficulty a future Labour Government would have renationalising BNOC.

The government did remove some of the privileges BNOC had enjoyed under the Labour Administration but BNOC was not opposed to this as these privileges had already achieved their purpose in that BNOC was able to borrow on the open financial markets. BNOC's right to a 51 per cent equity holding in all new exploration acreage was removed and BNOC would in future have to apply for blocks on equal terms with private companies. BNOC was thought to have too many licence obligations on the UKCS and thus the government, in disposing of some of BNOC's exploration interests, was putting its faith in the private oil companies to continue the exploitation of North Sea oil. Consequently, BNOC advertised for offers for farm-ins on twenty-three of its licensed blocks.

The government had already ended BNOC's right of first refusal when oilfields changed ownership and BNOC was to lose its access to cheap financing through the NOA and also its sole right to licences outside normal licensing rounds was abolished. Furthermore, BNOC lost its right to sit on the operating committees of oilfields in which it did not have an equity interest.[64] Not unexpectedly the government also terminated BNOC's exemption from PRT and its advisory rôle to the Government.

The Department of Energy was generally in favour of these changes and would have benefited from them in that there would be a degree of transfer of responsibility from BNOC to the Department. BNOC itself did little to oppose these changes although Lord Kearton consistently maintained that it was unwise to depend on foreign multi-national oil companies to develop the UKCS oilfields. The attitude of BNOC was that the commercial advantages had now become no longer necessary. BNOC would be almost on an equal footing with private oil companies and because some changes were inevitable, BNOC's compliance would create goodwill with the Conservatives. Although BNOC lost its requirement to advise the government on oil matters, there remained a close liaison between BNOC and the Department of Energy. BNOC retained its key functions as the government's agent for implementing participation agreements. The actual changes that had come about were neither as far-reaching as the Conservatives had wanted nor did they occur with any notable speed.

In the autumn of 1981 David Howell was replaced by Nigel Lawson as Secretary of State for Energy. A major priority for Mr Lawson, who has been described as being 'as zealous an advocate of privatisation as is to be found in the present (October 1982) Cabinet'[65] was the denationalisation of BNOC. Privatisation was a central part of overall Conservative policy and fundamental to Conservative Party ideology. The delay in dealing with BNOC was increasingly causing political embarrassment to the Conservative Government both publicly and within the Party itself. It was of considerable political importance to the government to privatise BNOC and although most privileges given to the Corporation by the previous Labour Administration had been withdrawn the principle of privatisation had not been achieved. Furthermore, the burden of BNOC's proposed expenditure (£2000m by 1986)[66] would, at least partly, have to be met by the Treasury; thus the Government were keen for BNOC to acquire private-sector financing.

The Chairman of BNOC, Mr Philip Shelbourne (appointed in May 1980 to replace Lord Kearton's successor, Mr R. Utiger) was also in favour of the privatisation of BNOC. Thus with the zeal of Mr Lawson backed by the Prime Minister, the Treasury and Mr Shelbourne, new impetus was given to the Conservative's commitment to privatise BNOC. The Oil and Gas (Enterprise) Act, 1982[67] provided powers for the government to split off BNOC's upstream activities into a subsidiary company which could then be sold, in part, to the general public. Ownership of the exploration and production company, Britoil, was transferred to the Secretary of State for Energy on November 1982. On 10 November, 51 per cent of Britoil's issued share capital was offered for sale on the Stock Exchange. In December 1982 Britoil became a private sector oil company with a government shareholding of 49 per cent.

BNOC remained solely as an oil trading company with functions to 'secure and dispose of UKCS petroleum in a way which contributes to national security of supply . . . to ensure that the UK economy receives the maximum benefits from the Corporation's access to such petroleum . . . and to act as the government's agent in the sale of oil taken as royalty-in-kind'.[68] Thus the government retained considerable and direct influence in the UK oil market. The political value of the state controlled BNOC was too great for the government to lose.

BNOC's practical use was soon to be emphasised in the political arena following OPEC's London Meeting in March 1983. The government was able, through BNOC, to price North Sea oil at such a level as tacitly to support the OPEC pricing agreement and prevent OPEC members (specifically Nigeria) from attempting to undermine the oil price deal.[69]

The government intended to raise almost £550m from the sale of Britoil shares, plus an £88m debenture repayment. The method of sale of Britoil shares decided upon was through a tender, with incentives to encourage participation from 'small' investors. The decision for a tender was based largely on political reasoning resulting from criticism of the government following the under-pricing of Amersham International. Similarly, various incentives given to 'small' investors (those applying for under 2000 shares) were largely due to political considerations. The government was eager to be seen to encourage 'the people of Britain to take a direct personal stake in the North Sea'.[70] An additional consideration was that the existence of large numbers of small investors may deter a future Labour Government from re-nationalising Britoil.[71]

Four main incentives were given to 'small' investors in the sale of Britoil shares. First, small investors could opt to buy shares at the 'striking price' rather than specifying a bid price. Second, like large investors, those applying for under 2000 shares could pay in two stages, £1 per share on application and the remainder on 6 April 1983. Third, a 'loyalty bonus' of one free share for every ten held and kept for three years was provided for small investors. Fourth, a simplified application form was made available in main Post Offices and High Street banks.

Due largely to the timing of the share sale (when there were widely reported[72] prospects of a weakening in the world oil price) a week after the Stock Market launch of Britoil shares, they were trading at 81p on the part-paid £1.[73] In October 1981 it had been hoped that the Government could raise between £750m and £1.2b by the sale of BNOC's production interests.[74] On 22 November 1982 Nigel Lawson was able to announce 'Britoil has now been successfully privatised on eminently fair terms for the taxpayer'.[75] However, the small investor had not responded in great numbers to the government's encouragement to purchase Britoil shares. Although Britoil was free from uncertain political and bureaucratic interference, the government retained certain powers within the company. As long as the government shareholding in Britoil is greater than 20 per cent the Secretary of State for Energy is empowered to appoint two directors to the Board of Britoil. In addition, the government possesses the special 'golden' share which may be used to control any change in ownership of Britoil. The special share was designed to prevent ownership of Britoil passing to foreign interests.

The delay in the privatisation of BNOC illustrates how government oil policy intentions can be deflected by short-term political considerations. The privatisation programme was central to Conservative policy and the privatisation of BNOC was an important political symbol of the Conservatives' commitment to free enterprise. Mr Lawson was able to achieve considerable personal stature in the Conservative Government by implementing the privatisation policy thus enhancing his career prospects within the Party. Following the Conservative's General Election win in June 1983, Mr Lawson was promoted to become Chancellor of the Exchequer. The personal competence of Mr Lawson, as Energy Secretary, with the backing of the Prime Minister, was an important factor in the privatisation of BNOC. The existence of the private sector company, Britoil, showed that an important political objective of the Conservative Government had been achieved. However, a 'ratchet effect' of government policy may be seen. The Conservative Government retained a 49 per cent shareholding in Britoil as well as the special powers associated with the 'golden' share. In addition, the government ensured it had the potential for control (with respect to directing and pricing substantial quantities of North Sea oil) through its ownership of BNOC as an oil trading company. The Conservatives were unable or more likely, unwilling to restore the initial policy situation with regard to participation and control (i.e. the situation existing prior to the 1974 Labour Administration) with no direct involvement in the UK oil industry.

The abolition of BNOC announced in March 1985 seems to have been a result of the continual criticism of BNOC being employed to intervene in the market by the Tory Government. More importantly, the losses made by BNOC during the autumn and winter of 1984–5 (due to BNOC having to sell large quantities of oil at the spot price having bought it at the higher, official price) caused the government considerable embarrassment.

6 Depletion Policy

6.1 INTRODUCTION

The concept of a government depletion policy, by definition implies that the government is unwilling to allow producer companies to determine their own production profiles. The implicit assumption thus being made by the government is that it is superior in its ability (relative to the industry) to identify some optimal production profile and it is capable of implementing policies which will result in this objective being fulfilled. The overall objective of this chapter is to examine the validity of these assertions in practice.

The case for a UK government oil depletion policy is examined in the context of government pronouncements and actions since the 1964 Continental Shelf Act. As in previous chapters, the concepts of the economic theories of politics and bureaucracies are employed in analysing the reasoning behind the government's formulation and adoption of a depletion policy.

Following an introductory description of government attitudes to depletion policy up to 1974, there is a brief outline of the economic case for government intervention in the oil industry to secure an optimal resource depletion rate in the North Sea. Government aims are analysed and the economic and political considerations underlying the arguments for a depletion policy are highlighted. The third section attempts to explain government behaviour since 1974 in the context of a political and bureaucratic framework. The fourth section examines the period 1974–80 and finally the major alternative policy instruments are considered. By identifying the pressures on the government policy process it is possible to examine the development of a depletion policy in the UK which is 'imperfect'[1] and to suggest future policy trends in the framework of an imperfect government.

6.2 THE HISTORICAL BACKGROUND (1964–74)

The decade prior to the 1973–4 world oil price increases was characterised by a lack of government intervention in the UK offshore oil industry (see Chapter 2). Both major political parties were keen to encourage the rapid exploitation of any resources that might be found on the UKCS. Thus relative to most other oil producing regions there was an attractive fiscal régime, generous licensing terms and an absence of explicit depletion controls.[2] In the mid-1960s there was considerable uncertainty as to the potential size of any UKCS oil reserves. After having discovered gas in the southern basin of the North Sea, oil companies' expectations of finding oil were high. Expectations of future price and cost trends during the mid-1960s were affected by forces working in opposing directions. As exploration activity increased in more hostile environments, costs were expected to rise and as world demand increased this would tend to force prices up. However, these pressures were expected to be offset by the rapid development and internationalisation of new technology, competition between oil companies, cheap Middle Eastern supplies, and shale oil.

Successive governments during this period created an environment in which the oil companies would be willing to commit vast resources for long periods of time into extremely risky ventures. Although at this time there was no oil discovered and therefore to deplete, governments' ambitions of rapid exploitation did apply to the future extraction of oil. Governments had concluded that, "the balance of advantage to the UK lay in exploiting and extracting these reserves of gas and oil as quickly as possible".[3] This policy was nevertheless backed by legislation[4] which gave the Energy Secretary significant potential powers with regard to introducing depletion controls.

By the beginning of the 1970s the UK government began to modify its rapid exploitation policy in response to a changing world oil market. The publication of the Club of Rome paper, *Limits to Growth*,[5] the shift in power away from the western multinational oil companies towards the oil producing states and the discovery of sizeable reserves in the northern North Sea, all acted to alter the popular perception of the international oil market. Up to this time UK depletion policy had not been an important political issue, nor had there been serious questioning of the government's intention to promote rapid exploration and also rapid extraction. Presumably this policy meant the government intended to allow the oil companies to determine their own production profiles according to commercial criteria, but it could be interpreted as meaning that the government was prepared to speed up production to some rate above that preferred by the oil companies.

When substantial reserves of oil were discovered in the North Sea the government began to re-examine its offshore oil policy. In common with other oil producing regions, once the government became aware of the potential economic and political value of its offshore oil a gradual trend towards increasing its revenue from, and control of, North Sea operations became apparent. Part of this overall policy trend was the explicit introduction of discussions of depletion controls into the general debate on offshore oil policy. The reasons for the timing of this development are central to the formulation of domestic oil depletion policy (and to oil policy in general).

6.3 THE POLITICAL RATIONALE FOR A DEPLETION POLICY

As has been explained in Chapters 2 and 3, by 1974 North Sea oil had become a major political issue. Numerous national and international factors combined at this time to make it politically necessary for all the political parties to highlight their domestic oil policies. Before examining the effects of various bureaucratic and political pressures on the depletion policy process it is necessary to outline the theoretical case for government-imposed depletion controls. The case for government depletion controls in theory is based on the assertion that an imperfect industry is unlikely to bring about an optimal depletion rate.[6] However, the oligopolistic oil industry does not necessarily deplete resources 'too fast'. Because many environmental costs of oil production are not internalised and are therefore not accounted for in oil prices, there is a tendency towards a 'too rapid' rate of depletion. In other words, if oil prices are 'too low' due to the presence of externalities not included in the oil price, then consumers are likely to demand greater amounts of oil than if the oil price reflected the precise costs involved. However, there is an important factor to set against the tendency for resources to be depleted too fast as a consequence of neglected environmental costs. Oligopolies tend to price higher, and fix output lower than in a competitive market situation resulting in a tendency to deplete oil 'too slowly'.

Secondly, the assertion that market interest rates are higher than the social time preference rate, and that therefore producers extract oil at a faster rate than society desires, similarly does not necessarily cause oil resources to be depleted 'too fast'. High interest rates in the economy tend to depress investment generally, so that in the oil industry (with low rates of economic growth) there will be a slow-down in the rates of depletion.

It is important to note that it is very difficult to determine the socially desirable rate of depletion. In addition, it is clear that there are many problems associated with determining whether the actual rate of depletion is 'too fast' or 'too slow'. It is accepted that the UK offshore oil industry is characterised by many imperfections but that in itself does not justify government-imposed depletion controls.[7] The following analysis attempts to highlight imperfections in the government depletion policy process which detract from the government's ability to intervene effectively in the oil industry, even if the market rate of depletion would be non-optimal.

The UK sector of the North Sea had become, by the early 1970s, a region with proven oil reserves. The first (and relatively small) discovery of oil on the UKCS was as early as 1967.[8] However, it was the discovery of the Forties Field in 1970 (estimated total recoverable reserves of 261 million tonnes) and Brent Field (estimated total recoverable reserves of 219 million tonnes)[9] which began to draw public attention to the wealth creating potential of North Sea oil. Thus the success of the private oil companies – at this stage in finding oil – effected the commencement of a gradual change in the public's consciousness with regard to North Sea oil. The public began to expect to be informed as to how the government would handle the development of the oil reserves and it was at this time that official announcements as to the possibility of government imposed depletion controls began to be reported in the Press.[10] The Committee of Public Accounts in 1973 received evidence from the Department of Trade and Industry who for the first time suggested that the government was considering the possibility of phasing out its rapid exploitation policy and that they could foresee circumstances in which there would be an 'advantage in delaying the exploitation'[11] of North Sea oil reserves.

The international oil market during the winter of 1973–4 was subjected to considerable upheaval and confusion. The OPEC oil embargo and the subsequent quadrupling of the price of oil thrust UK domestic oil policy to the forefront of public awareness. Media coverage of the 'oil crisis' was extensive and had the effect of focusing attention on the activities of oil companies (many of which were foreign-owned multi-national corporations) operating in the North Sea and in the rôle of the government in 'safeguarding' the nation's natural resources. With respect to depletion policy, the events in the

world oil market of 1973–4 were of considerable and specific importance. On 1 October 1973, the crude marker, Arabian Light, was
priced at $3.011 per barrel f.o.b. By 1 January 1974, the price was
$11.651 per barrel f.o.b.[12] This price increase had a significant effect
both on the economics of North Sea oil and on the UK Government's
attitude towards it.

In July 1974, the Labour Government published an oil and gas
policy White Paper[13] which attracted attention chiefly due to the
Government's proposals concerning state participation in the North
Sea and the establishment of BNOC (see Chapter 5). The Labour
Government stated that it would 'take power to control the level of
production in the national interest'.[14] Although the Government was
keen to encourage exploration and development it felt that in the
future depletion controls would become necessary and therefore the
powers needed to impose controls should be established as soon as
possible. These announcements were made at the same time as more
controversial participation plans, but declaration of the acceptance in
principle of the need for depletion controls was not a contentious
issue in itself. The need for depletion controls was accepted by all the
major political parties, although there were differences as to the
method by which any controls should be implemented. For example,
the Scottish Nationalists (who at that time were very important to
national politics) favoured a slowdown of development plans and a
production ceiling of 50 million tonnes per annum.[15] The Conservatives
argued for an Oil Conservation Authority to control production.[16]

This agreement can be explained by various political and economic
factors, and is crucial to the analysis of this chapter. The central
thesis of Downs' economic theory of politics[17] states that politicians
(and political parties) adopt policies to maximise votes and not
necessarily because of the desirability of the outcome of the policy. In
the context of depletion policy this is closely related to Breton's[18]
concept of an individual voter's 'coercion threshold' (see Chapter 3).
Downs maintains that the constitutional structure of a representative
democracy acts to remove those in power from direct contact with
voting citizens and knowledge of their preferences and wants. Thus
any hint of interest in, or concern for, an issue that politicians can
clearly identify is likely to be seized upon and developed into the basis
of some policy. The media attention bestowed on the 1973–4 international 'oil crisis' and its implications for the UK's indigenous
reserves thus acted to raise the public consciousness with regard to
North Sea oil, and also to lower the individual's coercion threshold.

The public had become relatively well informed on the subject and the cost to the individual in obtaining this information had been minimal. It thus became a political necessity for political parties to formulate oil policies.

Furthermore, Downs argues[19] that politicians adopt policies in order to hold office because of the intrinsic rewards of holding office. This concept can be extended, in a two party system, to include Tullock's analysis[20] of the effects of inter-party competition on the adoption of a policy by a political party. There is a desire by the two major political parties to adopt a policy that will not allow the opposing party to adopt a more popular policy. As there is already a perception of need for a depletion policy, both parties are keen to formulate a policy which will be at least as popular as the opposing party's policy. Due to the lack of explicit knowledge the parties have of individual voter's preferences, their policies will tend to converge on each other. Caution results from the lack of accurate information in the possession of the political parties and also the high cost (in terms of lost votes) of developing and adopting a 'wrong' policy. It has been postulated[21] that the incumbent party is able to gather a greater amount of information – which is of better quality – than the Opposition is able to accumulate, due to its greater resources (i.e. in the Civil Service). This does not necessarily result in the incumbent party always adopting the 'best' (in terms of vote maximisation) policy but it is quite possible for the 'wrong' policy to be adopted by both parties. If, for example, the Department of Energy advises the government to employ some depletion policy because the Department will benefit from the existence of that policy then the government, dependent on the Department for information and advice, may formulate a policy based on that advice. The Opposition, assuming the government is able to identify the 'correct' policy, copies it. The result is that both parties may adopt a 'wrong' policy. Thus there will always be an element of 'policy imitation' by one party if it feels the other party is better able to gauge public opinion on an issue and this will again tend towards both parties adopting similar policies.

In the case of depletion policy both major political parties have accepted in principle the need for government at least to have the ability to intervene and control production. The method by which this is to take place, although unclear in detail, differs between parties. This difference can be explained, to an extent, by Downsian type analysis whereby politicians are constrained in their attempts simply to maximise votes by intra-party rivalry, ideology, macro-political priorities and external pressures. These factors prevent the political parties formulating a policy with the sole purpose of maxi-mising votes. Added to this is the 'fog of uncertainty'[22] in which political advisers function which makes the vote maximising policy itself unclear; thus the probability of both parties adopting 'wrong' policies is relatively high. This can lead to the establishment of a third party which feels it can identify voter preferences more accurately.

Because the major political parties receive similar information with regard to what is politically required in order to maximise votes there tends to be a degree of agreement in principle – in this case for a depletion policy. The implication is that political parties may adopt policies that are contrary to party ideology because of the perception of popular support for that policy. In lacking a discernible policy the political party risks the possibility of allowing the opposing party to capture an unacceptable number of votes. In the context of present day depletion policy it would be expected that on the basis of ideology the Conservatives would tend to dismiss the idea of a depletion policy as it is interventionist. However, due to perceptions of political need the Conservatives have accepted the existence of the machinery for the implementation of a depletion policy in order to maximise votes. But, because of pressure from ideologues within the party, they are reluctant to enforce the policy as wholeheartedly as a Labour Government. Similarly, Labour's commitment to control has also been modified (see below) due to different governmental con-straints.

The agreement between the parties on the principle of depletion control can partly be explained by political expedience but also by an agreement on the economic factors[23] which combine to make a de-pletion policy desirable in economic terms. Whether these economic considerations (for example, security of supply, self-sufficiency, flatten-ing the production 'hump', etc.) are based on mistaken assumptions is not at issue at this point. What is relevant is that these factors were commonly perceived to be the economic reality and decisions were made from that basis.

Following the 1973–4 oil price increases the world oil market has been characterised by considerable uncertainty. This has resulted in the belief that the security of supply of North Sea oil is of great strategic importance to the domestic economy. Related to this idea is that, following the 1973–4 oil 'crisis', political parties reached the conclusion that it would be economically and politically undesirable for the domestic economy to be disrupted to such a degree in the event of another oil embargo or supply interruption. Political pressure to be seen to adopt some policy which could prevent such disorder was made more intense by the well-publicised 'excess' profits which accrued to the multi-national oil companies as a result of the world oil price increases of 1973–4.

There was also a general acceptance that oil prices would tend to increase over time, thus making it economically advantageous to invest in oil in the ground rather than extract it in the present form. This argument was enforced by the belief that oil company discount rates were high (relative to the government's) and therefore apparently placed too little emphasis on the long term. Again, it is not the issue here whether or not these are valid assumptions; they were the perceived reality from which basis decisions were formulated.

These factors together with increased estimates of North Sea oil reserves, led to an acceptance of the opinion that self-sufficiency in oil should be maintained for as long a period as possible and that there should then be a 'flattening of the hump' of domestic oil production.[24] There is also a popular belief that oil should be treated differently from other traded commodities because of its strategic importance to the economy.

The above factors directly influence the evolution and development of a government depletion policy and are political and economic signals which are received by both major parties. Due to these pressures on the formulation of government policy both parties have reached a certain consensus on the principle of depletion policy. There are further factors which indirectly influence political parties' attitudes towards depletion policy and these are more general points which apply not only to depletion policy but to oil policy in general. The overall trend of increasing government intervention in the North Sea oil industry since 1974 can, at least partly, be explained by the economic theory of politics and the theory of bureaucracy. A popular belief has been that North Sea oil is too important to be left to the control of foreign oil companies whose interests may act against some sort of 'national interest'. The result has been that political parties can maximise votes by being seen to be adopting measures to control the activities of the oil companies and to oversee the exploitation of North Sea oil, ensuring the 'national interest', however defined, is protected. Niskanen-type analysis would support this pressure for increasing intervention because the Department of Energy would rationally attempt to expand in size and budget; therefore it would attempt to influence the government so that certain policies emphasising certain characteristics would be adopted. The government depends on the Department of Energy for a considerable proportion of its information with regard to the domestic oil industry. The incumbent party, whether Labour or Conservative, would be subject to these very strong pressures in addition to the attraction, in political terms, of being seen to act to protect a 'valuable national resource'. Government intervention can tend to be self-perpetuating because of bureaucrats expanding their budgets and because of their opposition to attempts to reduce their budgets. Thus it would seem to be harder for a Conservative Government to enter office and reduce intervention than for a Labour Government to increase intervention. A characteristic of Mrs Thatcher's government has been the commitment to keeping departmental expenditure down. Thus it is possible for a Minister to achieve prestige and success by reducing, or at least controlling, the budget size of his department. However, faced with opposition from senior bureaucrats this has proved to be a difficult task (see Chapter 5).

6.4 GOVERNMENT BEHAVIOUR 1974–1980

In the early 1970s the government began to further the extent of its control and influence in the North Sea and also to tighten the forms of licensing policy (Chapter 4) and the fiscal régime (Chapter 7). One aspect of this trend was the consensus opinion on the need for a depletion policy. Although there has been agreement on the need for a depletion policy there has been a significant absence of explicit pronouncements and activity with respect to depletion control.

The 1975 Petroleum and Submarine Pipelines Act[25] which legislated to create BNOC also confers on the Energy Secretary extensive powers with respect to depletion controls. There was uncertainty with regard to the precise intentions of the government and as a result a set of guidelines, the 'Varley Guidelines' (see Chapter 2) were announced.[26] These guidelines were designed to reassure the oil companies as to the extent that the government would exercise its powers of depletion controls and its introduction of further powers. But still the government's intentions were far from clear and the Energy Secretary had virtually unlimited powers to control depletion in the vague interest of the 'national good'. Oil companies could do little to contest his rulings. Pressure and lobbying from the oil companies had the effect of the government revising or withdrawing some sections of the Bill during its passage through Parliament. Parts of the Bill which empowered the Minister to determine maximum and minimum production rates were withdrawn. However, the Energy Secretary still possessed extensive powers in the area of production rates. Producers were instructed to submit programmes detailing their capital investment plans and maximum and minimum production rates for oil and gas. The Energy Secretary had the power to reject programmes if they were not in the 'national interest' or were considered not to be consistent with 'good oilfield practice'. Both these criteria can be interpreted so as to encompass almost any conceivable circumstance that the Minister may arbitrarily include. Furthermore, if the programme is approved by the Energy Secretary, he can give the producer a 'limitation notice' which specifies limits within which the Minister can, by using a further notice issued after some designated period, direct the companies to produce. Another notice may require either a speeding-up of the producer's production plans in the case of a 'national emergency', or a slowing down, in order to advance the 'national interest'. Although the producers have the benefit of pre-specified production limits it is up to the discretion of the Minister as to how to define what exactly constitutes a 'national emergency' or the situation which requires intervention in the 'national interest'.[27] Furthermore, the 1976 Energy Act gives the Minister for Energy further powers to control oil and gas depletion and it is uncertain as to whether the 'Varley Guidelines' apply to controls exerted through the Energy Act rather than the Petroleum and Submarine Pipelines Act.[28]

With respect to government depletion policy, the Petroleum and Submarine Pipelines Act and the 'Varley Guidelines' are consistent with the Downsian and Niskanen-type analyses outlined previously. Although the depletion policy intentions, as stated in the Petroleum and Submarine Pipelines Act, are the policy of Labour politicians as a single entity within the government, there are pressure groups and factions within the Labour Party and elsewhere in the policy process. Decision-makers at Cabinet level, the Energy Secretary and junior Ministers, would be heavily reliant on the civil servants in the Department of Energy for information and guidance. Obviously there are political and ideological constraints imposed on the Department of Energy outside which their recommendations would be unacceptable to a Labour Government and there would be the views of other Ministers and political factions within the parliamentary and national parties to consider. However there are very powerful groups within the Labour Party who would be sympathetic towards policies that attempted to increase the influence and control of the government in the economy generally. The bureaucrat in the Department of Energy would attempt to expand the budget size of his department (by expanding its capacity for control in the North Sea and for the introduction of further controls in the future), thus fulfilling his maximand by supporting policies which accord with Labour Party ambitions with regard to the rôle of the state.

Therefore the Department of Energy, keen to increase its sphere of influence in the North Sea, might well support the type of depletion 'policy' outlined in the Petroleum and Submarine Pipelines Act and the 1976 Energy Act. The Energy Secretary, because of the extensive powers granted to him by these Acts, would be heavily dependent on the Department of Energy for advice and information. This would not only give the Department of Energy a continuous duty to consider depletion policy other than for more general purposes of regulation – but the Department would also be required constantly to update and revise plans for the oil sector and the energy sector as a whole.

Breton gives examples of what he calls 'typically bureaucratic behaviour'[29] and it can be seen that the structure of depletion policy put forward in the Petroleum and Submarine Pipelines Act fulfils many of the properties that government bureaucrats desire in a policy. These characteristics of policy include the introduction of elaborate machinery, systematically redefining the objectives and purpose of a policy to ensure it remains up-to-date and favouring economic planning. It should be noted however, that the Department of Energy itself is not a monolithic structure with a single set of ambitions. Divisions within the Department may have contradicting objectives; for example, the Oil Division has the closest contact with the oil companies and thus may tend to favour the company viewpoint and prefer fewer detailed controls. The Energy Secretary would view the extensive powers given to him in the Petroleum and Submarine Pipelines Act as increasing his personal stature, and his bargaining power within the government would be strengthened. Thus by expanding the influence (and budget) of his department the Energy Secretary is indirectly enhancing his own 'ranking' on the government front bench. This trend would have been gathering momentum as energy – oil specifically – became of significant political and strategic importance in the 1970s. Thus the Energy Secretary could exploit this trend by taking the opportunity to widen his sphere of control.

Reaction to these depletion plans from oil companies was, at this stage, somewhat muted. The oil companies were prepared to accept the principle that the government needs the legal powers to control depletion rates in extreme circumstances, but their main concerns were the extent of the controls and the possibility of arbitrary implementation. This, as with BNOC's originally proposed functions, highlights a major difficulty faced by the government in developing policies to control the North Sea oil industry. In legislation of this kind, which attempts to construct the legal framework for government intervention at some time in the future, there is an intrinsic difficulty relating to industry confidence. The world oil industry is characterised by considerable uncertainty and the UK government, in attempting to legislate in order to be able to react to any potential upheaval in the oil industry, can add considerably to that uncertainty. The desire on the part of the Department of Energy for flexibility[30] in order to safeguard domestic oil supplies in the event of some national or international disruption conflicts with the industry's need for relative stability and certainty. This 'security of supply' argument is often put forward by the Department of Energy as a reason for government control, or at least the power for control, in the offshore oil industry. However, this argument is similar to the 'national interest' argument in that both are vague and imprecise terms which are popularly very appealing and thus, with respect to the politician, can ensure vote maximisation. With respect to the bureaucrat within the Department of Energy, a flexible policy based on a politically attractive maxim gives the department a great deal of discretionary power and thus helps fulfil the maximisation of budget size objective.

It is not in dispute that the central government must possess the legal and institutional capabilities to intervene in the offshore oil industry in the case of a genuine national emergency. However, the system of control outlined in the Petroleum and Submarine Pipelines Act (for example, to alter production rates in the 'national interest') seem superfluous to the necessary requirements of the government to respond to a large-scale upheaval in the oil market. The 'Varley Guidelines' were specifically designed to assure the oil companies that the government would not alter production programmes outside the limits laid down. This did little to placate the fears of the oil companies (although it did postpone the implementation of any depletion measures until 1982 at the earliest)[31] as the Minister's extensive discretionary powers remained.

The Labour Government's depletion plans published in the Petroleum and Submarine Pipelines Act were somewhat overshadowed by the inclusion in the same document of the government's plans with regard to participation and the establishment of BNOC (see Chapter 5). These plans received a great deal of public attention because of the radical nature of the proposals. The Conservative Opposition and the private sector oil companies criticised the Act overwhelmingly on its plans for participation, but the plans for depletion controls seemed to be given the tacit approval of the Opposition. This was largely because the Conservatives recognised the political need for the existence of a depletion policy even though their precise method of implementation might have been different.

Thus the 1974–9 Labour Government constructed the legal framework for the introduction of a depletion policy although, as far as can be known, they did not implement any aspects of it with respect to oil. Following the election of Mrs Thatcher's Conservative Administration in 1979, there was apparently little change in the government's attitude to oil depletion policy. A conspicuously non-interventionist Conservative Government could, somewhat naïvely, be expected to reduce state control in the oil sector as a matter of political principle. The Conservatives made it clear that they intended[32] to dismantle BNOC and reduce state control in the North Sea. However, the Conservative Government continued with the existing policy which effectively was one of 'no-policy', i.e. the Conservatives had a very similar perception to the Labour Government as to the political – in vote maximising terms – value of a depletion policy. It was not until July 1980 that the Energy Secretary made an explicit policy statement[33] on depletion policy.

In this statement it was made clear that the government's objectives were to 'prolong the high levels of UKCS production to the end of the century'.[34] This was to be achieved by increasing exploration and at the same time enforcing some kind of depletion control. Again the precise method of depletion control was not yet clear, although Mr Howell, the Energy Secretary, stated that both production cutbacks and development delays were being considered (possibly with respect to the Clyde and 'T' Block projects). The overall content of the statement was such that the government did little more than express its very general intentions with respect to depletion policy. The statement outlined the government's overall attitude to depletion in that it publicly emphasised the government's concern over depletion policy and demonstrated that the depletion debate within government was continuing. Thus the statement fulfilled an important political function in that although Mr Howell did not outline specific policy proposals, he made it clear that the government considered the concept of government-controlled depletion to be an important aspect of energy policy. This retained public support and attention and avoided giving the Opposition an opportunity to capture votes by formulating its own policy. It also highlights the quandary facing successive UK governments since the mid-1970s in that the government appreciated the political necessity for announcing the existence of a depletion policy but appeared to be unable to formulate a policy consistent with other economic objectives which satisfied the ambitions of the powerful groups within the government which help develop the policy. This was especially difficult for the Conservatives because of ideological opposition within the Party to controls. The statement illustrates the difficulty the government experiences in attempting to reconcile various interested groups, and it is possible to identify references to some of these groups in the statement.

The justification given by the Minister for a depletion policy throughout this statement is overwhelmingly political. The reasons presented are that vague general conditions will be fulfilled, i.e. the 'national interest', 'security of supply' and 'good oilfield practice'. It has been emphasised that all these terms, whilst being popularly attractive, unless explicity defined are somewhat nebulous. They are open to considerable interpretation which effectively places a great deal of discretionary power with the Minister, allowing almost any policy to be subsequently developed and justified as desirable on the basis of its being consistent with the 'national interest'. Similarly, unless the government gives a detailed explanation and definition of what it considers to be 'strategic security of supply grounds'[35] this can also be used as a justification for almost any future intervention. To an extent the Minister is also keeping all options for future controls open, whilst at the same time ensuring popular political support by referring to 'optimum oil and gas recovery in the national interest'. Although 'optimum oil and gas recovery' has precise economic implications, unless these are made explicit it can again be interpreted in many ways. Because the 'optimum recovery' is to take place 'in the national interest' the likelihood is that bureaucratic and political objectives may supersede economic considerations.

The bureaucratic pressures within the political system are evident from the references to the need for firstly, a 'flexible approach'[37] and secondly, for continuous 'supervision' and scrutinisation of licence applications.[38] These functions are both consistent with bureaucratic ambitions for maximising budget size. The Minister is thus careful to consider the views of the powerful pressure groups within the government policy process, and also the overall objectives of the Conservative Party and Government. Viewed in this context it is perhaps not surprising that the Conservatives' attitude towards depletion was, under Mr Howell, largely a continuation of the previous Labour Government's attitude.

6.5 A POLICY OF 'NO-POLICY' – 1980 AND THE FUTURE

A fundamental problem confronted in any discussion on depletion policy is to identify clearly what is actually being discussed, i.e. what is meant by the term 'depletion'. Related to this definitional problem is the issue of the purpose of a depletion policy. If groups or individuals involved in discussions on depletion have an interest in the outcome of the debate, then it is possible for them to bias the discussion towards their objectives.

Since the mid-1970s, debate on the depletion of UKCS oil and gas resources has been primarily concerned with the downward control of production or, more specifically, with a flattening out of the so-called 'hump' once expected in the mid-1980s. Depletion policy, however, is about the rate of extraction of a non-renewable resource and thus it is not axiomatic that it should be concerned with the reduction of the rate of extraction. Recent discussion concerning depletion policy has focused on reducing extraction rates in the North Sea but this, as pointed out to the Energy Select Committee is a 'narrow interpretation'[39] of depletion which fails to consider associated issues. Thus it often happens that when depletion is discussed, the related issue of 'repletion' is neglected.

In the area of depletion, oil companies are primarily concerned with a two-fold objective; 'depletion of existing reserves and repletion of those resources as they are depleted'.[40] In the physical extraction of oil, the operating companies possess a great deal of technical expertise to which the Department of Energy does not have direct access. However, from the evidence presented over a period of eight months to the Energy Select Committee on North Sea depletion policy in 1981–2 it seems that the secondary aspect of repletion has been given little attention by the Department of Energy, with discussions concentrating on the relatively narrow issues of reducing production levels or delaying development.

A strict definition of depletion is important because, to an extent, it determines the nature and direction of the whole debate. The desires of government bureaucrats and political advisers for a depletion policy may implicitly bias and prejudice the depletion debate. Moreover, if oil companies expect the government to introduce some sort of depletion policy, the evidence may be formulated on the basis of the government having decided, according to its own political criteria, on the introduction of a depletion policy largely irrespective of oil company opinion. Thus there may be elements of strategic behaviour by the oil companies when giving evidence to the Energy Select Committee, perhaps by concentrating on how a depletion policy might be implemented rather than the merits of a depletion policy *per se*. The discretionary licensing system may be used to secure oil industry co-operation by the government (see Chapter 4) and oil companies may be reluctant to lose the goodwill of the government by arguing too forcefully against the principle of a depletion policy.

The Select Committee on Energy Report[41] on depletion policy emphasised this need to include repletion in the debate. The Committee suggested there is a need to maintain and to encourage exploration in the North Sea into the 1990s as by this means greater knowledge of the resource base can be gained. Without a high degree of knowledge of the existence and location of oil reserves (with respect to commercial accessibility) on the UKCS, it is very difficult to discuss depletion proposals in detail. Restrictions on production in the short and medium term can cause detrimental longer-term effects with regard to production and knowledge.[42] Thus the overall discussion of depletion needs to include aspects of policy designed at least to encourage the most complete exploratory activity on the UKCS as well as narrower aspects of reducing production or delaying development. The government is in a position to create a commercial and economic climate which does not inhibit activity on the UKCS by not, for example, introducing disincentives to investment in the oil tax system (Chapter 7) or by not introducing further political uncertainties to the oil industry (Chapter 6).

Lack of information and knowledge is a serious problem in the introduction of a depletion policy. Inherent uncertainties in the oil sector arise out of a lack of knowledge of the size of potential reserves, the rate of production per time period, and the price of oil. Thus, even before any instrument of depletion is explicitly considered, the uncertainties involved are already very extensive and cast doubt on the government's ability to alter depletion rates to achieve an optimal outcome – even if one was theoretically obtainable. In examining the possible instruments of depletion control, further uncertainties and inconsistencies can be highlighted within the policy process.

Production cutbacks and development delays are the two most obvious instruments for government action to reduce the rate of depletion. Although there is broad agreement on the side of the oil companies as to the net effects of government depletion controls, there is a divergence of opinion concerning which of the two methods is preferable. As far as the government is concerned, the main advantage of production cutbacks over development delays is that production cutbacks can have an immediate impact and are more susceptible to 'fine-tuning'. Nevertheless, from the politicians' viewpoint, production cutbacks postpone revenues which would have accrued in the immediate future and are therefore politically undesirable (especially to the Treasury). Development delays postpone revenues which would not have appeared for a period of years and are thus less important to the politicians now in office. Furthermore, all forms of downward controls risk passing on revenues to a government of another political party.

The oil companies' main objections to production cutbacks arise from their implications for company plans and finances (although this effect can be reduced if the timing and extent of cutbacks are known in advance). Plans made by companies would necessarily consider predicted cash flows and if production was subsequently disrupted, this could cause serious cash flow problems for oil companies. In addition, inefficiencies could result from idle capacity which oil companies could, in the short term, do little to offset. Although Esso were adamant that their 'interests are not served by production cutbacks'[43] they maintained that when compared to the likely alternative of development delays their preference was for production cutbacks. Because of the very long lead times in the oil industry, immense problems of forecasting exist with respect to the implementation of development delays. By examining depletion policy in the context of the economic theories of politics and bureaucracies, it is possible to see how the government justifies imposing development delays when in doing so it implicitly assumes that it is superior in its ability accurately to forecast prices, cost and demand conditions in the 1990s. The government is also assuming that the oil companies, in response to market signals, will not react by altering their production profiles, irrespective of government directives. Once the development delay has been announced there can be little flexibility due to the long lead times. Idle capacity may accrue, but not necessarily in the form of underused rigs or other capital equipment, but in the form of highly skilled human capital. Expert teams of geologists and engineers may, for instance, be split up or redeployed away from the North Sea. These factors, cumulatively, could result in a serious loss of momentum with respect to the exploitation of North Sea oil. Moreover, fears of further possible intervention would mean that oil companies would not be keen to come back into the North Sea after they have started to phase out their operations on the UKCS.

Shell favoured development delays rather than production cut-backs because they are the least disruptive in the sense that the development costs would not be incurred in the first place.[44] Esso, however, argued[45] that the loss of momentum of exploration and development where there are development delays outweighs the drawbacks of production cutbacks. This difference within the oil industry may not be significant given Shell and Esso's overall attitude to depletion controls. It is misleading to consider each potential depletion instrument in isolation as direct depletion controls are inter-related with other aspects of North Sea oil policy which have an indirect effect on depletion.

Shell's preference for development delays is argued alongside a criticism of the licensing system. Although licensing is not explicitly employed as a tool of depletion control, the nature of the discretionary system is such that the Department of Energy can offer certain blocks, or accept certain projects, on the basis of very general criteria. More importantly, the licensing rounds have been very irregular in both size and frequency which has impaired the ability of oil companies to plan into the future (see Chapter 4). The Department of Energy favours a discretionary system of licensing and oil companies do not object to this system. Oil companies operating in the UK sector of the North Sea have learnt how to use the system effectively and there are also bureaucratic tendencies within oil companies which would favour a discretionary system (see Chapter 4). With a more formal structure to licensing (such as yearly rounds as proposed by Mobil and Esso, or two-yearly rounds as proposed by

Shell[46] and with special terms to promote the development of frontier areas)[47] plans could be made with a greater degree of certainty. Thus the total effect of various government policies in the North Sea (which create an overall environment) is what the oil companies attempt to change rather than the effects of one policy in isolation. In the 1960s and early 1970s, the British Gas Corporation was used as an instrument of gas depletion policy. A re-nationalised BNOC could be used by a future Labour Government to control depletion. Thus the existence of policies which indirectly affect depletion add to the discretionary powers of the Department of Energy and also may be employed by government bureaucrats in tactical bargaining with politicians, other departments and with the oil industry.

A similar argument can be put forward with respect to taxation and how it relates to depletion. The Department of Energy is consistent and explicit in emphasising that the fiscal régime in the North Sea is designed to have a neutral rôle with regard to depletion.[48] Although the oil tax system may not have been intended as a depletion control device, there has nevertheless been an impact on oil companies' expectations and this has resulted in plans being revised or shelved. The tax system in the North Sea, since the introduction of Petroleum Revenue Tax, has been subject to continual change and alteration (see Chapter 7). This has again resulted in considerable uncertainty. Whether bureaucrats within oil companies would actually prefer a relatively simple, stable tax system, is not at all certain. But a system which allows the oil industry 'to plan its long-term investment programme with reasonable confidence'[49] would, *ceteris paribus*, in terms of economic efficiency, be desirable. Secondly, because PRT is not based solely on excess profits it does not work as a progressive tax (Chapter 7). This can act as a disincentive to the development of marginal fields (with long lead times and high risk) where taxed profits are greater than excess profits. It has been reported[50] that oilfield development projects have been shelved (the Tern Field) evidently because of dissatisfaction with the tax system. Although these decisions could be tactical manoeuvres by oil companies (in order to exert pressure on the government to alter the oil tax system), the result is nevertheless a delay in development; i.e. an effect on the rate of depletion.

Thus the evidence presented to the Energy Select Committee by oil companies was not solely concerned with direct measures to influence depletion. The oil companies took the opportunity to voice their dissatisfaction with the overall policy of the government towards

North Sea oil. If depletion is defined in its broad sense as including repletion and encouraging the acquisition of knowledge of the resource base, then oil taxation and licensing policy do have an impact on depletion. This reasoning is consistent with the conclusions made by the Energy Select Committee whose main criticisms of government policy are of licensing[51] and oil taxation.[52]

Other possible instruments of depletion control such as royalty banking or keeping BNOC equity oil underground, can be seen as unlikely methods to be implemented. This is not because they are inefficient or impractical, but because by employing these methods the full costs (in terms of revenue foregone in the short term by the Exchequer) fall on the government or a state corporation. As it is the government that is implementing the policy, it seems doubtful that it would be sympathetic towards these methods. Equally, the oil companies would favour them, although it would involve the re-nationalisation of Britoil and an increased political rôle for the national oil corporation, to which oil companies would be strongly opposed. The Energy Select Committee's Report maintains that royalty banking is the preferable method of depletion control in the short term as if 'intervention is deemed essential in the national interest . . . it is right for the Government to bear the risks and that royalty banking is therefore preferable to production cutbacks'.[53] If the government is able to forecast that the rate of world oil price increases would in the future exceed its own discount rate, the government would be better off keeping its oil underground. Thus it could employ a method of royalty banking and profit in doing so. However, the oil companies, who probably possess superior expertise, would do this anyway without government directives.[54]

Thus within the area of government depletion policy, there are many differing pressures exerted on the policy process. The 'national interest' objective is interpreted by various groups within the policy process so as, at least, not to conflict with that group's ambitions and if possible, actively to further that group's objectives. This phenomenon, highlighted by the rôle adopted by the Treasury, seriously undermines the theoretical case for government depletion controls. The Treasury and the Department of Energy are adamant that policy decisions of this nature are made at Cabinet level and that in advising their respective Ministers and providing them with information, the two departments are equal partners.[55] Given that the overriding objective of the Conservative Government's macroeconomic policies is to reduce the rate of inflation, a major consideration is the control of public expenditure. In this context Cabinet decisions would be more inclined to be sympathetic to Treasury recommendations which would be made with regard to maintaining government oil revenues whilst the Department of Energy might be keener for some sort of oil conservation policy. The strength of bargaining power of the Treasury can, at least in part, explain why no oil depletion policy has as yet been implemented. With respect to the postponement of BNOC's (now Britoil) Clyde development programme,[56] this can also be viewed not as a Department of Energy inspired oil conservation measure, but an example of the Treasury's reluctance to provide finance for the project in the short term. This emphasises two factors with respect to government oil policy which are especially significant to depletion policy. First, the government is prepared to subjugate a 'national interest' depletion objective it considers may exist in response to other economic objectives. Also, the government's time horizons do not necessarily seem to be longer than those of private sector oil companies, although far-sightedness is often given as a reason for government intervention. Second, inter-departmental rivalry and bargaining are important in 'guiding' ministerial decision-making. It seems the present Administration's depletion policy of 'no-policy' stems from its being consistent with the Government priority of controlling inflation and with its non-interventionist beliefs.

The Treasury, in its opposition to a depletion policy that would diminish its North Sea oil monies (and especially to the method of royalty banking), is behaving consistently with Niskanen-type theory (Chapter 3). This also seems to be the case with respect to other groups active in the policy process such as oil companies, politicians and government bureaucrats. Each group is attempting to steer government policy so as to protect or further its self-interest objectives. Changing economic and political circumstances constantly alter the relative bargaining power of these groups (for example, a forceful personality as Energy Secretary replacing a weaker one) and therefore create uncertainty as the policy is continuously subject to review and modification.

These uncertainties inherent in the government policy process are in addition to the vast inherent uncertainties of the oil sector. It seems doubtful whether the Department of Energy is able to forecast more accurately than the oil companies. Even if it could, it is not certain that its recommendations would be implemented if they clashed with Treasury objectives, or even with other objectives of the

Department of Energy. Oil companies regard depletion as an integral part of government oil policy in general and their major concern is the overall effect of oil policy on their activities. For this reason, oil companies concentrated their criticism, in evidence to the Energy Select Committee, not on depletion control but on oil taxation. In June 1982 a further minor change was made to the oil tax system followed by more major changes in March 1983 (see Chapter 7). This demonstrates that constant pressure exerted on the government has some effect, and that government policy is a constantly changing process.

Given the various different interests within the policy process it seems unlikely that the present Government will implement a direct depletion policy. An expedient solution (as the Energy Select Committee recommends)[57] would be some sort of extension of the 'Varley Guidelines' with a great deal of discretionary power remaining with the Department of Energy and the Energy Secretary. On 8 June 1982, Nigel Lawson – then Secretary of State for Energy – announced in the House of Commons that production cutbacks would not be imposed before the end of 1984.[58] Thus, effectively, the existing 'policy' has been extended for a further two and a half years. This enables senior politicians to maximise votes by appearing to be making decisions about depletion rates, the Treasury retains its short-term revenues and the Department of Energy keeps its extensive functions of acquiring, up-dating and analysing information from the offshore oil industry. In other words, an extensive apparatus of depletion control is in being but its effect on oil company decision-making is probably minimal, thus reflecting the unresolved conflict between a non-interventionist government and a budget-maximising department. Government attempts to flatten a 'hump' of production can only add to the already present uncertainties in the oil sector and at the very minimum, knowledge of the size of the 'hump' (and if it still exists)[59] is necessary prior to any debate on the desirability of a depletion policy. Even if the government can accumulate knowledge of UKCS oil resources, uncertainties within the policy process, as outlined above, significantly distort policy proposals. When highlighting all the 'imperfections' involved in formulating and implementing depletion policy, it becomes apparent that the effects of such a policy are likely to be quite different from what they would be under the perfect government of welfare economics.

7 Taxation Policy

7.1 INTRODUCTION

The following chapter attempts to analyse the development of North Sea oil taxation since the early 1970s. A complicated tax structure has evolved and the emphasis of this chapter is on the reasons why the system has become so complex and why it has changed so frequently, rather than on the effects on company profitability or government revenues of the various tax rates and allowances. The economic theories of politics and bureaucracies are applied to the changing tax system in order to provide some explanation of the rationale behind the changes. The extent of the changes is considerable. For example, Petroleum Revenue Tax (PRT) was introduced in 1975 at a rate of 45 per cent. Since then, it has been increased to 60 per cent, 70 per cent and currently is 75 per cent. A Supplementary Petroleum Duty (SPD) was introduced in 1981 and subsequently abolished in 1982, Advance PRT (APRT) was introduced in 1980 and is now to be phased out, and the tax-free oil allowance after being halved, is now to be restored to its original level for fields granted development consent after 1 April 1982. Many more changes in the tax system have been made. By highlighting the main pressures and influences on the governmental policy process an understanding of the directions of likely pressures and the relative strength of these pressures may be acquired.

The first section of the chapter concentrates on the First Report from the Committee of Public Accounts, Session 1972–3[1] (hereafter the PAC Report). This influential document was highly critical of the existing tax system and one of its proposals was for the introduction of a specific tax on revenues from oil production. Following the publicity surrounding the publication of the PAC Report and the world oil price increases of 1973–4, the Labour Government introduced the basic tax structure for the offshore oil industry in the 1975 Oil Taxation Act (OTA). The impact of the OTA is examined in the second section. The 1975 OTA is crucial to the analysis of this chapter because it sets down the original taxation framework in the light of which all the later proposals and recommendations occur. The third section is concerned with the present Conservative Government's record as regards oil taxation. This has been characterised by numerous modifications and amendments to the system.

Various pressures from within government bureaucracies and political parties can be seen to have affected the development of the tax system. Extraneous factors such as changes in the world price of oil are similarly important, as are perceptions of instability and expectations of change. Industry pressure groups and independent analysts also have their rôles in influencing the policy process. These and other factors have contributed to the formation of the present day North Sea oil taxation system.

The final section attempts to summarise these effects and draws attention to the uncertainties and inefficiencies created by a constantly changing tax régime.

7.2 THE PAC REPORT

As previously stated (Chapter 2) two factors in the early 1970s combined to make North Sea oil a major political issue. The discovery of sizeable oil reserves on the UKCS from 1970 onwards and the fourfold increase in the world price of oil in the winter of 1973–4 necessitated the formulation of comprehensive domestic oil policies by all the major political parties. Both these factors were necessary conditions for the politicisation of North Sea oil. If sizeable oil reserves had been discovered and world oil prices had not risen as they did in 1973–4, or if oil had been found in significantly smaller quantities and the world price of oil had increased substantially, then in both cases it is unlikely that political parties would have considered North Sea oil a sensitive vote-capturing area.

Prior to 1973, concern with the tax system[2] was focused on the practical functioning of Corporation Tax and Royalties. In the 1960s and the early 1970s governments in the UK were keen to create an attractive environment which would encourage foreign oil companies to commit resources to the UK sector of the North Sea. In the early 1970s once significant reserves had been found the 'rapid exploitation'[3] policy remained. Governments were still attempting to encourage oil companies to invest in the North Sea and were reluctant to impose conditions which might firstly act as a disincentive to the companies and secondly, might lead to OPEC retaliation in some form. The PAC Report found these considerations to be exaggerated in their importance.[4]

The system of taxation was strongly criticised in the PAC Report on the basis that Corporation Tax, as it functioned, had many loopholes and anomalies which could result in future tax revenues from the North Sea being unacceptably low. The Committee of Public Accounts maintained that, compared to other oil producing countries, the taxes imposed by the UK Government were low and company profits would be higher than elsewhere.[5]

The PAC Report noted[6] that tax losses (of around £1.5 billion for the nine major companies) available to offset Corporation Tax had so far accrued and losses could continue indefinitely. This potential loss to the UK Exchequer had resulted chiefly from two factors. Firstly, oil companies in the Middle East used a posted price for tax calculations and this price had become greater than the market price. Because oil companies used this posted price as a transfer price, the companies' tax profits on oil production tended to be inflated, whilst profits from their downstream activities in oil consuming countries were correspondingly understated. Trading losses for tax purposes thus accumulated in the UK shielding the companies from Corporation Tax on their downstream activities. Upstream, the Exchequer did not receive tax revenues from company profit because the tax liability could be offset against the payment of overseas tax.[7]

Further criticism of the tax system was that a company operating and making profits in the North Sea could offset these profits against its other activities outside the North Sea.[8] The PAC maintain in their Report that these 'artificial tax losses'[9] arising out of loopholes in the system were unacceptable in terms of the direct revenue lost to the Exchequer and also in terms of the harmful effect on the balance of payments.[10]

In their recommendations, the PAC advocated that considerable changes be made to the existing tax system in order to 'substantially improve the effective tax yield from operations on the Continental Shelf'.[11] The PAC recommended that changes be made which would prohibit UKCS monies being pre-empted by taxes elsewhere and also that capital allowances for extraneous activities should be controlled.[12] A further and very significant suggestion made by the PAC was for the introduction of a specific oil tax 'the Government . . . will consider among other methods the possibility of imposing a system of quantity taxation, e.g. a barrelage tax'.[13]

Evidence received by the PAC highlighted fundamental shortcomings of the system of government control. Witnesses from the Department of Trade and Industry (DTI) maintained that under the existing taxation régime substantial revenue would accrue to the Government.[14] At the same time, the Inland Revenue estimated that the revenue likely to accrue to the Government up to 1980 would be, due to the method of transfer pricing and capital allowances, significantly less than the DTI estimates, and would probably be negligible.[15] Discrepancies such as this illustrate a lack of communication between government departments which inevitably detracts from the efficient functioning of government. Niskanen-type analysis (see Chapter 3) suggests that government departments may withhold (or select) information from the sponsor, i.e. the politicians, in order that the bureaucrats' utility function may be maximised. However, with different government departments performing overlapping functions, the departments may tend to withhold information from each other in order to maintain their exclusive, and therefore crucial, expertise. Thus planning decisions may be made by departments with very imperfect knowledge or with inaccurate information. On this basis, senior Civil Servants advise and inform Ministers and political decision-makers. The PAC noted that the DTI lacked information (specifically with respect to costs)[16] but this is a criticism not so much of a lack of communication between departments as of a lack of co-operation between government departments and offshore industry.

The PAC Report itself was criticised by the industry[17] because no industry witnesses were invited to submit evidence to the Committee. This lack of communication between a parliamentary body and the offshore industry resulted in the industry not being able to explain the tax system from its standpoint. It has been argued[18] that had the industry been invited to present evidence to the Committee it would have been able to contest the Inland Revenue evidence and would also have outlined important cost expectations.[19]

The recommendations of the PAC implied considerable criticism of the government departments responsible for the North Sea oil tax system. Nevertheless, the types of changes favoured by the PAC would have been broadly consistent with government bureaucrats' ambitions. Extensive modification to the existing tax system as well as the introduction of an oil tax would mean an expansion of departamental work and responsibility. The PAC Report did seem to have an impact on the policy-process;[20] however world oil price increases during the winter of 1973–4 pre-empted any government action based on the PAC Report. In 1974 the vote-maximising potential of North Sea policy became the priority consideration of the government.

This change of emphasis can be analysed within the framework of the theory of bureaucracy and the economics of politics. Government bureaucrats in the Treasury and the Inland Revenue attempt, within certain constraints, to maximise taxation revenue from any specific industry.It is possible that the bureaucracy's desire for taxation revenue, in order to expand its budget size, is greater than the politicians' desire for a specific sum to be raised to meet specific expenditure plans. Whilst the politician attempts to extract revenue in order to finance certain economic and financial projects, the bureau may be simply attempting to maximise taxation revenue constrained only by what it is practically able to extract.[21] This distinction becomes less important in times when politicians' economic policies enforce the raising of considerable sums by the Government; both political and bureaucratic ambitions would then tend to coincide. Prior to the 1973–4 price increases the amount of potential taxation revenue that could be captured from the offshore oil industry, within political and economic constraints, was not being reached.[22]

The term 'taxable capacity' refers here to that amount of taxation revenue the government could appropriate from a given industry or sector. The 'taxable capacity' of an industry is determined by various political and economic factors and the perception of these factors by the voting public and by politicians and bureaucrats. Thus prior to 1973–4, the 'taxable capacity' of the offshore oil industry, as perceived by the PAC, was approximately determined by the taxation paid by oil companies in other oil producing regions.[23] This amount was not expected to be captured by the existing taxation system. The PAC proposals were designed to ensure that government tax revenues were approximately equal to the oil industry's 'taxable capacity'. The world oil price increases of 1973–4 had the effect of increasing the 'taxable capacity' of the offshore oil industry. The perceptions of the world oil market was such that the government, by reacting to this general perception, could increase its oil tax revenue. Oil companies operating in the North Sea were expected to earn 'windfall' profits. UKCS oil became strategically very important to the domestic economy and it was necessary for political parties to develop comprehensive oil policies. The popular attitude towards the oil companies operating in the North Sea was generally unsympathetic and the government was under considerable political pressure to ensure that it had close control of oil company activities and avoided their earning 'excess' profits.

7.3 THE 1975 OIL TAXATION ACT

Thus the 1974 Labour Government was keen to act swiftly with regards the oil taxation system. The Government's White Paper[24] presented to Parliament in July 1974 outlined the Government's overall intentions with respect to North Sea oil policy (Chapter 2). The government intended to take measures in order to 'secure a fairer share of profits for the nation' and to 'assert greater public control'.[25] The taxation proposals fell into two categories. First, the White Paper outlined technical changes to Corporation Tax so that loopholes highlighted by the PAC Report would be closed. Artificial losses arising out of the multi-national oil companies' transfer pricing policies would be eliminated and a ring fence would be constructed in order that 'receipts from the North Sea should not be at the mercy of allowances and losses resulting from extraneous activities'.[26]

Secondly, the White Paper proposed 'an additional tax on the companies' profits from the Continental Shelf'.[27] This was intended as a specific tax on companies operating on the UKCS designed to recapture economic rent transferred to the oil companies via the discretionary licensing system (see Chapter 4). The government's intentions towards North Sea oil as stated in the 1974 White Paper were to control North Sea activities and to increase the Government's tax take. In the context of the economic theory of politics, the proposed oil tax was one important aspect of an overall oil policy, the development of which had, by 1974, become a priority government objective.

In November 1974, an Oil Taxation Bill (OTB) was published which outlined the new oil tax, the Petroleum Revenue Tax (PRT) as it was originally conceived. Although both the 1974 White Paper and the 1974 OTB contained many of the recommendations of the 1973 PAC Report the PAC's tentative suggestion of a barrelage tax was not considered viable. A barrelage tax, a tax similar to the $12\frac{1}{2}$ per cent production royalty, on the quantity of oil produced, was thought to be detrimental to marginal (low profit) fields as it would not take account of costs. Fears from the oil industry concerning some form of quantity tax were that since it was unrelated to profits the government would increase the tax if profits were to rise, whilst the reverse would not hold.[28] However, PRT did have elements of a barrelage tax in the sense that it was imposed on production revenues from each field.[29]

The form PRT took in the OTB was different from that made law in the subsequent 1975 Oil Taxation Act (OTA). As stated above, the government's overwhelming priority with respect to the North Sea oil tax system was essentially political. The government needed to be publicly seen to be taxing the 'uncovenanted'[30] profits of multi-national oil companies operating in the North Sea. During the report stage of the 1975 OTA negotiations took place between the government and the oil industry. In response to oil industry disquiet many changes were made to the 1974 OTB before it became an Act. The success of the oil industry, as a pressure group within the policy process, was at least partly a consequence of the uncertainties present in the offshore oil industry (for example, with respect to price and cost expectations and production estimates). The industry, in possession of expertise and information not available to the government, was in a relatively strong bargaining position.[31] However, the Government was very keen to pass an Act in order to maximise the vote-catching potential of offshore oil. This, to an extent, explains the haste (because the modified proposals concerning an oil allowance and an annual limit on tax payable, did not function as expected)[32] in which the modifications to the OTB took place.[33] The OTB was publicised on 19 November 1974 and on 25 February 1975 Mr Edmund Dell, then Paymaster-General, announced various modifications.

The structure of PRT as a flat rate tax remained in the OTA. Because the taxable unit was the field rather than the company, it was not possible for an oil company to offset an unattractive field against a more profitable field.[34] It was ostensibly the disincentive effect on marginal fields[35] of the originally proposed system to which oil companies were chiefly opposed. During discussions with the companies the government introduced various changes to the tax system designed to diminish any disincentive effects of its original proposals. The modifications outlined by the Paymaster-General[36] fall into two categories of discretionary and non-discretionary provisions.

First, the discretionary method granted the Secretary of State for Energy, with the consent of Treasury officials, powers to 'defer, waive or refund royalties in the whole or in part'.[37] Any refund would be exempt from Corporation Tax and PRT. The non-discretionary provisions included increasing the 'uplift' on capital expenditure for tax purposes from 50 per cent to 75 per cent, the introduction of an Oil Allowance and the introduction of a tapering and safeguard provision to limit PRT payments. The 75 per cent Uplift was designed to 'give the industry a further element of front-end loading that is free of PRT'.[38] Postponing the payment of PRT has a significant effect on development, since due to the discounting of cash flows, the early years of a field's life are very important in determining the commercial viability of that field.[39] The Oil Allowance was for half a million tonnes of oil per six-monthly period free of PRT, subject to a cumulative total of 10 million tonnes per field. Although this allowance was designed to benefit marginal fields it applied to all fields. The intention was to benefit smaller fields more than proportionately,[40] but with a production period so short that the cumulative total of 10 million tonnes was not reached (once the 75 per cent Uplift had been recovered) the benefit to some marginal fields might not be as significant as originally perceived.[41] The Safeguard provision provided a limit on PRT chargeable and the Tapering provision was to ensure that the PRT payment did not exceed a proportion of capital expenditure. Together, the Tapering and Safeguard provisions (generally accepted as one measure) limited the payment of PRT to 80 per cent of net revenue minus 30 per cent of cumulative capital expenditure. If this was less than the PRT calculated in the normal way then the Tapering and Safeguard limit applied. PRT was introduced at a 45 per cent rate in February 1975.

In addition, included in the OTA was a restriction on Corporation Tax. A 'ring-fence' around a company's North Sea operations was constructed. As explained above, the object of this measure was to ensure that Corporation Tax payable on profits from North Sea oil production would not be diminished by losses in other areas of a company's activities. The second proposal concerning Corporation Tax changes outlined in the 1974 White Paper, i.e. with respect to 'artificial' losses arising out of transfer pricing policies of multinational oil companies, was also implemented in the OTA.[42]

The swift passage of the OTB through the House of Commons and the characteristics of the subsequent Act emphasised many of the main features of the economic theories of bureaucracy and politics. The political and economic framework of North Sea oil had changed significantly between 1972 and 1975. In 1973–4 the national political scene was uncertain, political parties were keen to identify political issues which were high in the public consciousness; North Sea oil was thus an obvious target. As with depletion policy (see Chapter 6) there was a high degree of consensus as to the concept of PRT. PRT was an extremely complex system of taxation, it would be implemented by the Inland Revenue and overseen by the Treasury. Government offices in these departments, as with those in the Department of Energy, would be, *ceteris paribus*, in favour of a complex system as their expertise and understanding would be vital to their policy decision-making superiors. Government bureaucrats would expand their powers by developing a politically important and growing function. Furthermore, their positions would be protected by their possession of information and expertise about the tax system, making their bargaining position relative to the politicians (sponsors) stronger. Independent commentators and academics would also have difficulty in examining the system and providing a check because of its complexity and the imbalances of information.

Oil industry opposition to the OTB concentrated not on its complexity but on its stringency. This concern resulted in various amendments being incorporated in the OTB at committee stage. However, mainstream economic theory would predict that oil companies would prefer a simple, more straightforward tax régime in order that plans and forecasts could be made with greater certainty. This would apply at the most senior level in oil companies where policy decisions are taken and where simplicity would be advantageous. Similarly, within the Civil Service some senior bureaucrats may favour a simple tax system, for example to reduce uncertainty in overall economic planning. If the theories of bureaucracy are applied to the oil companies (and to their lobby organisations such as The United Kingdom Offshore Operators Association, UKOOA, and The Association of British Independent Oil Exploration Companies, BRINDEX) it becomes apparent that bureaucrats within oil companies are likely to favour a complex tax system for reasons similar to those of government bureaucrats. A complex and unstable tax system is in the common interest of tax experts in the Civil Service and in oil companies. Such complex tax systems may be manipulated by these tax experts to their advantage; moreover, such a system is good for employment and status. Hence, suggestions for capturing rent by auctioning licences (see Chapter 4) or for a simplified tax system are not well received by bureaucrats in government or in oil companies. In addition, officers within industry pressure groups attempt to justify their own positions, the existence of their organisations and their status within the industry. A complex tax system which is perpetually in a state of flux facilitates the fulfilment of these ambitions.

The Paymaster-General, Edmund Dell, maintained[43] that PRT would be received and adjusted over time. Constant up-dates and reviews of policies are consistent with ambitions of tax experts in the government and also in industry, since the bureaucrat is able to maintain his position and expand his budget over time. The speed at which the amendments to the OTB were introduced illustrates the desire of politicians to capture as much political advantage as possible from the OTA whilst North Sea oil was still at the forefront of popular awareness. It also clearly shows the unwillingness of government bureaucrats to abandon a 'bad' policy and instead to modify and amend the policy; again this in line with Breton's[44] bureaucratic behaviour characteristics (see Chapter 3). This is a recurring theme with regard to domestic oil taxation policy.

Agreement between the two major parties on the need for a comprehensive offshore tax system was enforced by official statements designed to assure the industry that the Government and the Opposition were concerned with the long term. The tax was intended to be 'a stable tax and not used as a short-term regulator'.[45] This sentiment was echoed by the Opposition spokesman, Patrick Jenkin: 'there is no intention here that this should be anything other than a stable tax, which will not be used for demand management purposes nor as a short-term regulator'.[46] In the context of the economics of politics these statements can be seen to be designed to capture votes. Both parties are perceived to give priority to long-term stability and the Conservative Opposition is popularly seen to be placing the long-term interests of the oil sector above party-politics by supporting the tax system.

Policy statements and proposals made in the mid-1970s can now be examined in the context of the historical development of domestic oil policy. The techniques of the theories of bureaucracy and politics may be employed to understand and explain this development; here with respect to the offshore oil tax system. For example, in a Treasury Press Release,[47] Edmund Dell maintained that 'the Government will stand ready to review and adjust the incidence of PRT in the event of a sustained and significant change in the price of oil in real terms'. This, seemingly, would be interpreted to mean that if oil prices were to rise in real terms, PRT would be increased.[48] However, because of successive Government's desires to finance rising public expenditure there would be expected to be a 'ratchet effect' with regard to PRT changes. As oil prices have risen in real terms, Governments have been prepared to increase the tax rate consistent with Mr Dell's statement. But when real oil prices fell (e.g. by 18 per cent between 1975 and 1978)[49] the reverse did not happen. The 1983 oil tax changes were partly a response to a real price fall but the ratchet effect was still there because the tax reduction only partially compensated for lower price expectations.

A further example of the predictive powers of the economics of politics and bureaucracies can be seen with respect to the discretionary safeguards included in the OTA. The Secretary of State for Energy was able to use discretionary powers to refund all or part of the royalty payments made to the Government. Royalty revenues were to be made available to BNOC via the National Oil Account and it would be expected that both the Government and BNOC would object to the investment capabilities of BNOC being constrained by royalty refunds.[50] Whilst BNOC was funded by the National Oil Account (prior to its privatisation – see Chapter 5) refunds of royalties to oil companies did not occur.

7.4 1978 TO THE PRESENT DAY

The 1975 oil taxation package was constructed at a time, it has been argued,[51] when there were considerable uncertainties as to the future profitability of North Sea oil operations. By August 1978 the Government view[52] was that 'though many uncertainties remain we are in a position to take stock and it is apparent that companies are obtaining very large profits from the natural resources of the nation. We believe that the public share of these profits can and should be increased.' The Labour Government proposed three fundamental alterations to the offshore tax system.

First, PRT would be raised from 45 per cent to 60 per cent; secondly, the Uplift on capital expenditure would be reduced from 75 per cent to 35 per cent; and thirdly, the oil allowance would be reduced from 1 million long tons per annum to $\frac{1}{2}$ million tonnes per annum.

These proposals were announced before the Iranian Revolution, before the Iranian oil workers' strike in October 1978 and before the world oil price increases of 1979–80. Nevertheless, these proposals can in part be seen as defensive measures. The 1975 oil taxation package was implemented at a time when oil prices were expected to fall in real terms.[53] Thus it can be assumed that had oil prices in 1975 been expected to rise, or at least remain constant in real terms, the basic rate of PRT would have been higher than 45 per cent. In 1978 it had become apparent that world oil prices, in real terms, were not going to decrease significantly as a long-term trend and would probably increase as demand had started to increase after the initial post 1973–4 decreases. Thus in terms of the domestic oil industry's 'taxable capacity', in 1975 the Government had been mistaken in its oil price forecasts and was therefore not extracting revenue up to this capacity. In 1978 oil companies were seen to be obtaining large profits from their North Sea oil operations,[54] and with a General Election imminent political parties were keen to maximise votes on clearly identifiable policy issues.

It is significant, in the context of the economics of politics, that the Conservative Opposition were in agreement (in August 1978) with the proposed tax changes.[55] It is difficult to predict how effective oil company opposition to these proposed changes would have been had the world oil price increases of 1979–80 not taken place. By the time the proposals had been included by the new Conservative Government in the 1979 Finance Act (in July), the spot price of the Saudi 'marker' was $33.13 per barrel (from $12.70 at the end of 1978) and the trend was sharply upwards.[56] Thus the oil company bargaining position would have been relatively weak. Strong public antipathy towards multi-national oil companies would have enabled the new Conservative Administration to increase oil taxation revenue with little effective opposition from the oil industry.

There was also minimal intra-party opposition to these measures, at least partly reflecting the overall economic needs of the Government to increase its revenue. The North Sea oil tax system could be employed, within certain political and economic constraints, to increase Government revenue thus acting as a means to an end and also as an end in itself in maximising votes on a major policy issue. The Iranian Revolution in 1979, the Gulf War in 1980 and the world oil price increases of 1979–80 resulted in North Sea oil becoming, as in 1973–4, a key political issue. Costs of information, because of constant media coverage, were low and individuals were relatively well informed. There would be popular pressure on the government to act, which in this instance would be in line with the government's existing plans for raising revenue.

Thus the tax changes made in July 1979 (which also included a change in licensing regulations so that royalty payments would be accelerated) were not in direct response to the 1979 world oil price increases but in response to a change in oil price expectations which occurred between 1975 and 1978. That oil prices rose significantly in the first half of 1979 would have strengthened the resolve of the Government to increase the tax take. It would also have made the passage of the legislation through Parliament easier due to popular support and the weakening of the oil companies' ability to pressurise the Government into modifying, or introducing concessions into, their proposals. The desire of the Government to increase its oil tax take in order to maximise votes by controlling the activities of multi-national oil companies and responding to the public perception of OPEC-inspired price increases was consistent with the Government's ability to implement the policy due to the perception of oil company 'windfall' profits.

In March 1980 the Chancellor of the Exchequer announced[57] that PRT would be increased to 70 per cent for the bi-annual chargeable period ending after 31 December 1979. In addition, advance payments of PRT would be due in the first two months of each chargeable period. The Chancellor maintained that these changes were in response to the world price of oil increasing 'dramatically' in the previous year, a change which 'has greatly favoured the oil companies' and 'greatly strengthened the industry's cash position'.[58] Thus, again in response to the perception of 'windfall' oil company profits, the Government increased the tax rate.

In November 1980 the Chancellor announced his intention to introduce a further tax on UK offshore oil.[59] This new tax, the Supplementary Petroleum Duty (SPD), and other changes to PRT were formally introduced in the 1981 Finance Act in much the same form as they were originally proposed in November 1980. In announcing SPD, and thus subjecting North Sea oil companies to a four tier tax system, the Chancellor invited suggestions and proposals for alternative tax systems which would leave the Government with a broadly similar tax take from offshore oil. SPD was to operate from January 1981 to June 1982 when the overall tax régime would be reviewed.

In his 1981 Budget, Sir Geoffrey Howe outlined the full changes to the North Sea tax system. SPD was set at 20 per cent of gross revenues less an annual allowance of one million tonnes. The tax would be collected monthly thus maximising the adverse effect on company cash flows[60] (and emphasising the Government's desires for short-term revenue). Moreover, two changes to PRT were announced. First, the 35 per cent Uplift on capital expenditure was to be restricted to the period up to PRT payback 'when an operator's cumulative incomings from a field exceed his cumulative outgoings'[61] (and later extended to include outstanding APRT liability). Previously, this Uplift applied to capital expenditure whenever it was incurred. Secondly, with respect to the tapering and safeguard provisions, they were now only to apply from the commencement of production for a length of time equal to 1.5 times the period from the start of production until PRT payback.

Thus, against the background of the oil price increases of 1979–80, the Government considered it possible to increase its tax take from North Sea oil operations. The Chancellor's invitation for proposals for a completely new tax system strengthened expectations as to future changes in the system. The changes to the Uplift and the tapering and safeguard provisions were largely in order to avoid the possibility of tax relief on capital expenditure reaching 100 per cent which would encourage inefficient capital expenditure. With respect to the tapering provision, with an increased rate of PRT (in the previous Budget from 60 per cent to 70 per cent), more fields would have been pushed into the tapering limit, possibly enabling the companies to receive more tax relief than the level of expenditure in any one year.

The main criticisms of these changes concerned SPD. Because SPD was based on gross revenues and was unrelated to profits, it was considered adversely to affect high cost fields.[62] UKOOA[63] considered that the net effect of these changes would cause serious cash flow problems to companies during the declining years of a field which could lead to premature abandonment. The tax changes were seen as being 'dominated by the need to procure extra revenue very quickly'[64] seemingly a view supported by the Government's subsequent willingness to abolish SPD. In terms of the economics of politics, the Government, by the introduction of SPD, had exceeded the perceived 'taxable capacity' of the offshore oil industry. In response to the Chancellor's invitation for proposals for alternative tax systems both the oil industry (e.g. UKOOA[65] and BRINDEX[66]) and independent analysts (e.g. the Institute for Fiscal Studies) recommended the abolition of SPD. Concerted criticism of the overall tax system, and specifically of SPD, combined with a significant slowdown in development activity during 1981 (due in part to falling world oil prices) and resulted in further tax changes being announced in the 1982 Budget.

The strength of the industry as a pressure group by 1982 had increased relative to the Government's position. In the 1982 Budget alterations to the tax system were announced and although 'the tax changes are no more than a tinkering within an unchanged tax structure'[68] the Government did abolish SPD and there was estimated to be a small reduction in tax payments over time.[69] The IFS proposals for a new tax system were found unacceptable by both Government and oil industry for reasons which seem explainable in terms of the economic theories of politics and bureaucracies. The IFS[70] proposed fundamental structural changes to the oil tax system. A Petroleum Profits Tax (PPT) would replace the existing four tier tax system and would be levied on a field by field basis with a 'ring-fence' around each field. 'PPT would be levied in three tiers, each related to successively higher rates of return on investment'[71] and thus would be a progressive tax with fields having a higher rate of return bearing a higher average tax rate. To accept this new tax system Government bureaucrats would be implicitly admitting their original policy had been 'wrong'. A simplified tax system might result in a loss of jobs in both government and industry and thus would be opposed on this basis. Tax experts in government and in oil companies had accumulated skill in manipulating the existing system and a new, simple tax system could diminish their relative expertise and thus their relative power and job security. The introduction of a completely new system would also reflect on the sponsors of the existing system (i.e. the Government and the relevant Ministers) and could result in the danger of losing votes to a competing policy.

The main proposals outlined in the 1982 Budget were that SPD be replaced by Advance PRT (APRT) and to compensate for the resultant lowering of Government take the rate of PRT was raised from 70 per cent to 75 per cent. APRT operates similarly to SPD in that it is a 20 per cent tax on gross revenue with an oil allowance of one million tonnes each year. However, APRT is allowable against PRT; thus it does not affect the total amount of PRT paid but only the timing of the payments. Timing of payments is an important consideration to the Government and illustrates its strong preference for receiving tax monies as soon as possible. A further measure proposed in the 1982 Budget was that advance payments of PRT were replaced by a system of spreading PRT payments on a monthly basis.

Industry criticism was immediate and minor adjustments were made to the Budget proposals. The main criticism that APRT adversely affects early cash flow (reducing the attractiveness of all fields) and that it should therefore be abolished, was not accepted by the Treasury. The first concession, however, was that fields would only incur APRT for five years. Secondly, for a field where PRT is not paid because profits are not great enough, instead of APRT being refunded at the end of the field's life in a lump sum (which amounts to a long-term interest free loan to the Treasury) the companies would be repaid after five years.

The major North Sea oil taxation changes announced in the 1983 Budget fell into two categories. Firstly, where development approval was granted after 1 April 1982 (except for onshore fields and Southern Basin fields) royalty payments were to be abolished. In addition, the oil allowance was to be restored to its 1978 level of 1 million tonnes per year subject to a cumulative limit of 10 million tonnes per field. Secondly, applying to all fields, APRT was to be phased out over a period of four years and abolished after 31 December 1986. Also, PRT relief could be claimed on all future exploration and appraisal expenditure and restrictions on PRT relief for shared assets was to be eliminated. These proposed tax changes were met with 'surprise and jubilation among the oil companies'[72] and were specifically designed to stimulate exploration and development activity. The Government's perception of the oil industry's 'taxable capacity' had been exceeded and the Government acted to ensure that activity on the UKCS would continue into the 1990s at a level acceptable to the Government. The lack of orders for the UK oil supply industry added to the government's concern of losing votes. It is not until the mid-1990s when most of the new generation of oilfields are producing oil that there becomes a significant divergence between Government oil revenues on the pre- and post-1983 Budget tax system, and if new discoveries come on-stream in the 1990s because of the tax change this will offset reductions in Government tax revenues.[73] In the context of past oil tax changes the 1983 Budget was significant as it introduced a new type of structural change. By differentiating between fields under production or development (old fields) and projects granted development consent after 1 April 1982 (new fields), the Government has set an important precedent. It is now able to treat the North Sea tax régime as two distinct systems.

On the more mature, old fields, production is relatively insensitive to oil price changes, thus in the future the government could increase tax takes from these fields. For new fields, however, where development is sensitive to oil prices, the tax system could remain unchanged. If oil prices strengthen and increase during the next decade the Government might consider that it is not extracting revenues to the 'taxable capacity' of the oil industry. Thus it would be expected that in the 1990s the Government may increase its take from the new fields.[74] The dual tax structure gives the Government further complexity of taxation policy, a factor consistent with tax specialists' ambitions both in government and in the oil industry. However, uncertainty concerning future tax changes (most importantly in the medium and long term) is unlikely to have diminished as a result of the 1983 Budget.

Even after the substantial and well-received, (at least by the oil industry) changes of 1983, tax stability was not achieved. Further changes were announced in the 1984 Budget. CT was to be reduced over a period of time from 52 per cent to 35 per cent and the CT first year capital allowance reduced from 100 per cent to 25 per cent by 1986.

In the 1985 Budget it was expected that tax conversions would be introduced in order to encourage enhanced oil recovery and other incremental investment. This did not occur largely, it would appear, due to the loss of short-term revenue to the Exchequer even though in the longer run tax revenues would increase and recoverable reserves would increase.

7.5 A CHANGING TAX SYSTEM

The encouragement to development which the Chancellor hoped to bring about by the 1982 tax package did not occur. Oil companies' strategic bargaining in response to the tax system may partly explain the well-publicised development delays such as the Tern Field[75] (by Shell and Esso). This illustrates the possibility of tactical behaviour which may result in non-optimal allocation of resources over time. Oil industry pressure, in the form of lobbying and tactical bargaining, was a significant factor in bringing about the 1983 Budget tax changes which gave an immediate and considerable boost to North Sea activity.

Oil companies now expect tax changes which alter the economics of a project after the investment decision has taken place. The constantly changing tax system adds considerably to the uncertainties surrounding offshore oil. However, the Government believes it is important to have a degree of flexibility in its tax policy in order to be able to respond to unforeseen shocks. Flexibility may be achieved through changes in rates of taxation or changes in various concessions and allowances but constant changes in the structure of the tax system have been seen to result in instability and arbitrary effects[76] on oilfield projects.

There are two ways in which, on the basis of rational expectations with respect to a changing tax system, fields may not be developed or development may be postponed. First, if the tax system is expected at some time in the future to become much more onerous. Second, if oil companies attempt to pre-empt an increase in the tax system by postponing or abandoning development plans. In addition, expectations of tax increases will also give an incentive to accelerate depletion of fields already in production in order to produce as much as possible while tax rates are relatively low.

The overall tax régime in the North Sea, as explained in this chapter, has been changed many times since 1975. The considerable instability of the system reflects its unsuitability as an oil tax. A simple tax system would be advantageous to industry (and Government) planners and it would also facilitate independent monitoring of the system. However, as has been argued in this chapter, a policy which is complex and is frequently amended is generally favoured by some bureaucrats within the Government, within oil company pressure groups and by certain tax experts within oil companies. Decision-makers in oil companies in many circumstances may favour simplicity and stability and some bureaucrats in Government (e.g. in the Department of Energy's Oil Division) may also be averse to an unstable system. However, the theories of bureaucracies suggest that in the policy process bureaucrats have numerous methods by which they can protect and further their interests (see Chapter 3). The relative power of bureaucrats whose self-interest is fulfilled by an unstable, complex system evidently outweighs that of the bureaucrats who may favour stability and simplicity.

Government bureaucrats' ability to influence the development of the oil tax system has been greater than with regards to other policy issues because the bureaucrats' ambitions have been broadly consistent with those of the politicians in government. Whilst the politicians are concerned with the taxation system as part of North Sea oil policy and with the effects of that policy, the concern of Government bureaucrats (e.g. in the Treasury, Inland Revenue and some parts of the Department of Energy) is overwhelmingly for the taxation package *per se*. Niskanen-type analysis (Chapter 3) suggests that bureaucrats specialising in oil taxation are likely to maximise their utility functions by acquiring characteristics of oil tax policy associated with the structure and implementation of the oil tax. Thus for these bureaucrats taxation of North Sea oil is an end in itself rather than a means towards an end.

The Government's desire to obtain short-term revenue as part of its broader economic objectives has resulted, since 1978, in annual changes to the tax system. These constant changes tend to distort the link between prices and the development of North Sea oil resources. If oil companies assume that as oil prices rise, even in nominal terms, the tax rate will increase, this can act as a disincentive to investment. Oil companies consider the possibilities of future losses due to oil prices falling against future gains due to oil prices rising. If the companies expect profits resulting from oil price increases to be negated by an increased tax rate this leaves only the possibility of losses if prices fall. This acts against the desirable economic effects of a price rise, i.e. that as oil prices rise, producers are encouraged to develop more costly fields.

In response to world oil price increases the Government has been seen to be taking action over a 'national asset' and at the same time has been seen to be controlling multi-national oil company activities in the North Sea. The 1981 tax changes apparently over-shot the industry's 'taxable capacity'. In 1981 the trend in world oil prices was downward, North Sea activity had slowed down considerably and industry opposition to the tax system was at its most vociferous. The overall effect of the 1982 Budget changes was to reduce the marginal tax rate very slightly from 90.28 per cent to 89.5 per cent,[77] but this did little to stimulate activity. Thus the 1983 Budget changes may be viewed in the context of activity in the North Sea and in oil-related industries being unacceptably sluggish. Considerable concessions applicable to 'new' projects announced in March 1983 had the desired effect of stimulating exploration and development activity (e.g. Shell's announcement of plans to spend £800 million a year to 1990)[78] and boosted industry confidence. The economic theories of politics and bureaucracy would suggest, however, that when these 'new' fields commence production the Government may tighten the fiscal régime considerably. It may be the case that a third category of oil taxation will be developed. A new, lenient tax systems for fields about to be developed in the 1990s, a harsher system for those commencing production in the 1990s, and another lenient system for mature fields nearing exhaustion in order to lengthen their production tail.

8   An Overview

8.1 INTRODUCTION

Preceding chapters attempt to identify some of the major problems associated with the development of various aspects of offshore oil policy in the UK. The economic theories of bureaucracies and politics, as outlined in Chapter 3, provide the theoretical and conceptual framework in which specific problems of UK oil policy are analysed.

By making assumptions as to the behaviour and motives of bureaucrats and politicians within the government policy process a positive evaluation of present policy may be achieved. Knowledge of the presence, direction and relative strength of influences on government policy contributes to an understanding of the policy process. Given certain assumptions as to future trends of key economic variables affecting North Sea Oil (for example, world oil prices and exchange rates) potential bureaucratic and political pressures can be recognised. Although forecasting bureaucratic and political pressures in isolation is speculative, when seen in the context of other economic factors potential bureaucratic and political pressures assist in the interpretation of policy and the determination of policy trends. At the very least it may be possible to rule out certain policy developments where bureaucratic or political interests exert a particularly strong influence.

In providing an overview of North Sea oil policy and government, the second section of this chapter briefly summarises some important points of the four aspects of oil policy discussed in previous chapters. The third section outlines some conceptual problems involved in applying the economic theories of politics and bureaucracies to North Sea oil policy.

A central theme throughout the analysis of North Sea oil policy in previous chapters is that distortions and pressures on the policy process caused by political and bureaucratic self-interest are problems inherent in the UK system of representative government. The fourth section brings together the various aspects of North Sea oil policy and argues that once the inherent problems of the government policy process are considered the case for extensive government intervention in the oil industry becomes doubtful.

The fifth and concluding section examines the general rôle of the government in the North Sea oil industry. Principles for effective intervention are established and suggestions for improving the policy process are outlined. Areas for further research are encountered in the final section. Most importantly these include the development of a characteristics approach to policy determination and changes to the policy process itself.

8.2 SUMMARY OF UK OIL POLICIES

Licensing Policy (Chapter 4)

The decision in 1964 to adopt a discretionary system of licence allocation has had a considerable impact on the government's policy towards North Sea oil. Frederick Erroll, the Conservative Minister for Power, in 1964 made a statement[1] outlining the criteria on which the government would base its allocation decisions. These criteria showed a strong element of nationalism which subsequently remained in all licensing rounds as political and bureaucratic considerations in licence allocation increased. In choosing a discretionary system in preference to an auction system the government bureaucrats were able to increase their administrative powers and scope of control. An important reason for the development of the offshore taxation system was the realisation[2] that the discretionary licensing system was enabling oil companies to capture economic rent from their North Sea operations. The oil tax régime was designed to recapture economic rent for the government.

Irregularity in the size and timing of licensing rounds adds to confusion and uncertainty in the oil industry. Licensing rounds have tended to be used by the government to speed-up or slow-down activity on the UKCS, often for political reasons or as a result of other economic objectives.

Nationalistic policies which favour domestic over foreign companies in licence allocation are politically attractive though for a country dependent on international trade, they are of doubtful economic value.

The discretionary licence system enables politicians and government bureaucrats to pursue their own objectives. In a discretionary system government bureaucrats gather information, continually appraise oil company activities and generally regulate the industry. These functions enable government bureaucrats to 'steer' policy and tend to result in the superoptimal size of government output (see Chapter 3). In contrast, an auction system utilises the price mechanism ensuring that the lowest cost producer is the recipient of the licence and at the same time, in a competitive system, captures economic rent for the government.

Participation Policy (Chapter 5)

The major reasons put forward for the creation of BNOC were that it would facilitate security of oil supplies, permit control over the disposal of North Sea oil and be an effective instrument by which a national oil policy could be implemented. However, the establishment of BNOC by a Labour Government was overwhelmingly a political action. Active and direct participation in North Sea activities enabled the Labour Government to fulfil a socialist policy and attract votes by protecting a 'national asset'. In addition, the Government was seen to be controlling multi-national oil company activities at a time when oil policy was at the forefront of public consciousness.

BNOC's terms of reference were not precise; a public corporation whose relations with the Government and with the industry were unclear had the effect of contributing to a lack of industry confidence in government oil policy. An example of the politicisation of North Sea oil policy creating instability in the oil industry may be seen by the Conservative Government's privatisation of the exploration and production section of BNOC in 1982. A policy which becomes an important aspect of a political party's platform is liable to be changed by a subsequent government keen on differentiating its policies. The Labour Opposition in 1983[3] pledged its commitment to the re-nationalisation of Britoil.

Moreover, the establishment of a national oil company added a further bureaucracy active in the policy process with its own ambitions and interests.

Depletion Policy (Chapter 6)

Successive governments in the UK have acted ambivalently towards offshore oil depletion policy. Whilst accepting the political importance of the existence of a depletion policy both the major parties, when in government, have felt unable to implement such a policy. The postponement of the Clyde development project was a special case in that BNOC would have used monies from Treasury finances. As a result of the Petroleum and Submarine Pipelines Act, 1975, extensive depletion controls are in existence and the Secretary of State for Energy has considerable personal powers with respect to oil depletion. These powers increase the Minister's authority and stature and similarly are consistent with bureaucratic objectives. It seems the major reason for the non-implementation of a depletion policy has little to do with the economic wisdom of such a policy but because of the short run revenues which would be foregone by the government if production controls were imposed on oil companies.

Both bureaucratic and political factors have influenced UK oil depletion policy and the result has been uncertainty as to future government plans. Assurances as to Government intentions concerning development delays and production cutbacks by Eric Varley (in 1974) and Nigel Lawson (1982) alleviate company planning problems only in the short term; the fundamental uncertainty surrounding government depletion policy and its precise form remains.

Taxation Policy (Chapter 7)

The oil taxation system in the UK is characterised by its complexity and instability. Since 1975 there have been numerous changes and modifications to the tax system. Government bureaucrats and tax experts in oil companies have ambitions which may be fulfilled by the existence and maintenance of a complex, changing tax system. Thus to certain bureaucrats the structure and method of enforcement of the tax system is the major consideration of policy rather than the effects of the tax policy on the oil industry. The Inland Revenue's Oil Taxation Office itself was created as a result of the 1975 Oil Taxation Act.

Discontinuity of tax policy has led to anomalous situations, specifically with regard to high-cost marginal fields and the erosion of early cash flow (due most importantly to SPD and APRT). Tax policy is a crucial element in the economics of oilfield development and the uncertainty created by a constantly changing system necessarily affects company expectations and harms their decision-making.

8.3 CONCEPTUAL PROBLEMS

In attempting to identify influences and pressures within the policy process major conceptual difficulties arise. The government policy process involves numerous groups and individuals whose ambitions are not necessarily consistent with political objectives or economic efficiency. Furthermore, government bureaucracies cannot be seen as single entities with a single set of priorities and ambitions. Within government departments there are smaller, specialised groups and sections. The interaction of these groups within the overall bureaucracy is an important source of bureaucratic strategic behaviour. Conflicts may arise within a government department, between departments and with bureaucracies in oil companies. In applying the economic theories of bureaucracies to North Sea oil policy, conceptual difficulties occur largely because of the nature of government bureaucracies in the UK.

First, government bureaucracies are perceived to operate at a distance from the public. There is a general lack of knowledge surrounding the precise functions and terms of reference of government bureaucracies. This accrues partly from a tradition of official secrecy and as a result of bureaucratic strategic behaviour. Lack of information and secrecy tend to protect government bureaucrats from outside criticism and accountability. In addition, the power and influence of the bureaucracy is protected because of its possession of superior information.

Secondly, the relationship between government bureaucrats and politicians is one in which the bureaucracy is intended to be an objective and apolitical adviser to senior politicians. However, government bureaucrats as individuals inevitably have political opinions and make political judgements. Bureaucrats advising Ministers may bias their recommendations to be consistent with personal opinions or viewpoints. Government bureaucrats advise and make decisions based on a set of conditions which are often unknown. One can speculate about what these conditions are, but the priority or weighting given to the various considerations is information kept strictly confidential by the bureaucrat. Because the criteria for bureaucratic decision-making are unknown there is no objective or benchmark to judge the success or failure of bureaucratic output against. Again, the bureaucrat is protected against criticism. Since 1979, the reformed Select Committee system has tended to make senior bureaucrats more accountable for their actions and has provided information concerning activities of bureaucrats. Nevertheless, official confidentiality may still be used as an argument to protect the positions of government bureaucrats. Furthermore, when a policy is the concern of more than one department, which is often the case, priority of interests between departments is difficult to ascertain.

A third conceptual problem involved in examining bureaucratic and political influences in the policy process, and a reason why these influences are often under-stated, arises from the public perception of the Government as an omniscient guardian, protecting all members of society. This can lead to a willingness to allow Civil Servants and senior politicians to decide which actions and policies are in the interests of society. If governmental officers are viewed as individuals attempting to maximise personal utility functions the perception that their behaviour is altruistic becomes very doubtful.[4] Bureaucratic and political pressures influencing policy are not immediately obvious and are difficult to quantify. Nevertheless, it is possible to identify sources of conflicts of interests and recognise their relative strength and their directional influence.

Traditional neo-classical economic theory tends to regard the government as a single, homogeneous structure. Thus, in response to market imperfections, the government may act in order to correct any imperfections and approach a Pareto efficient solution. This clearly will not occur unless there is support for such a policy from groups within the government policy process.

8.4 GOVERNMENT AND NORTH SEA OIL POLICY

The development of North Sea oil policy has been characterised by statements from politicians of major parties which have been designed to maximise votes on important policy issues. Policies have been presented as being in 'the national interest', 'protecting a national asset' and 'guaranteeing security of oil supplies'. These statements have little precise meaning because they can be arbitrarily interpreted in numerous ways; but they are politically appealing sentiments. A policy developed as a response to public perceptions of a 'crisis' and argued for as being in the 'national interest' may achieve the object of capturing votes. However as crises pass and public perceptions change the initial policy is also likely to change, so that a stable policy is unlikely. The policy, whilst satisfying political and bureaucratic ambitions, may have little to do with promoting economic efficiency.

The four aspects of North Sea oil policy outlined above are interrelated to a degree, notably the oil tax system and the licensing system. However, the four policies have developed separately, often controlled and implemented by different departments, and successive governments have tended to maintain this separation. Oil companies consider the effects of oil policy as a whole and thus are inclined to trade-off the effects of one policy against those of another policy. This has led to oil companies, and to a lesser extent governments, adopting tactical methods of bargaining and strategic behaviour. Oil companies negotiate or bargain with the government in order to change or modify policies. In return for the government, say, making tax concessions, the oil industry may attempt to speed-up development activity. The participation negotiations (see Chapter 5) highlight the extent of tactical bargaining that has occurred between government and industry concerning government policy. Tactical behaviour and strategic bargaining (outlined in Chapter 3) exist in an imperfect industry where there are bureaucracies within oil companies. However, because of the politicisation of North Sea oil, the extent and importance of tactical decision-making and strategic behaviour increase considerably as government intervention in the offshore oil industry increases. As oil companies attempt to coerce the government into making policy concessions the degree to which oil companies are able to make decisions and plans on commercial or economic criteria is reduced. North Sea oil policies have been subject to political pressures which oil companies must then include in their decision-making process. A fully integrated national energy policy would increase the possibilities of inefficiencies as the size and complexity of the policy increased. Similarly, uncertainty in the oil industry would increase as political and bureaucratic intervention increased.

As the tendency for inefficiencies arising from political and bureaucratic behaviour increases as the government's involvement in the oil industry increases, the converse is also likely to occur. With government's functions in the offshore oil industry restricted to some sort of 'overseer' or 'protector' rôle, the potential for bureaucratic and political distortions is limited. Clearly defined limits to government policy would at least partly control the bureaucratic ambitions of expanding budget-size leading to a self-perpetuating increase in the influence and power of the bureaucracy.

Evidence in preceding chapters suggests that differences in oil policy between Labour and Conservative Governments have been limited. In a two-party system of government, if one party identifies a policy with vote-maximising potential it is likely that the second party (for instance, due to inferior information) will adopt a similar policy. The danger of losing votes because of a 'wrong' policy is a major consideration of political parties. Political consensus (for instance, with respect to depletion policy) may at least partly be explained by economic theories of politics as outlined in Chapter 3.

A Conservative Government succeeding a Labour Administration may be expected to reduce state intervention in industry. With respect to the offshore oil industry, the Conservative's success in this area must be qualified. An incoming Secretary of State for Energy may not be eager to reduce the powers of his own department because in doing so he may also diminish his own standing in the government. Existing powers of control are likely to prove useful to the government in order to influence the oil industry. Personal powers granted to Ministers (e.g. by the Petroleum and Submarine Pipelines Act[5] with regards to depletion controls) are often favoured by government bureaucrats because of the potential for steering policy and the achievement of other bureaucratic ambitions. A weak or inexperienced Minister can thus increase uncertainty and speculation concerning policy expectations by hesitancy and indecision. The greater the extent of personal powers granted to the Minister the greater is the success of policy dependent on the personal ability and strength of that Minister. The delay in the privatisation of BNOC may be partly attributable to expedience, but also reflects on the skill and competence of the Secretary of State in controlling a government bureaucracy and implementing policy. An additional reason for the failure of Conservative Governments to reverse interventionist policies is the existence of a 'ratchet effect'. A Conservative Government may intend to reduce state intervention in the oil industry but because of bureaucratic opposition and tactical behaviour (Chapter 3) it is unable to restore the initial policy circumstances. Again, citing BNOC as an example; its privatisation in 1982 created Britoil but the trading sector of BNOC remained in public control.

8.5 THE RÔLE OF THE GOVERNMENT IN THE NORTH SEA

The existence of oil policies in the UK implies that the government is able to improve upon the situation of an imperfect oil market. For this position to be justified the government must be capable of fulfilling certain conditions[6] whereby government policies and the ability of government to implement those policies may be seen to achieve a preferable (to the public) outcome than a non-interventionist policy.

First, the government must be able to forecast the outcome of a policy of no intervention in the oil industry. If intervention already exists any changes in policy must also be considered by comparing the outcome resulting from policy changes to the outcome of no policy changes. Unless this condition can be achieved it is not possible to gauge whether a government policy will improve the existing situation. Due to characteristics of the international oil market, forecasting in the oil industry is an extremely uncertain process. The government is dependent on its Civil Service for gathering information and constructing forecasts of future trends in the oil market under various key assumptions. Intrinsic shortcomings of the government system, whereby bureaucrats may select or withhold information from politicians in order to steer or influence government decision-making can result in policy recommendations being accepted on the basis of bureaucratic self-interest rather than their being based on objective forecasts.

Secondly, the government must have a clear idea of the policy preferences of society in order that the government's perception of the 'national interest' is consistent with society's 'national interest' objectives. Downsian theory and other aspects of the economic theories of politics (Chapter 3) highlight the difficulties of politicians and political parties within the UK system of representative democracy being able to identify voter preferences. Individuals must have considerable information made freely available to them if they are to have clear policy preferences. Even then, because of distortions in the voting system, individuals may not be sufficiently motivated to indicate their preference and it is unlikely that voter preferences on a single issue (within a package of issues) will be revealed to political parties.

Thirdly, once 'national interest' objectives have been identified, specific oil policies must be developed which fulfil these objectives and are at least preferable to the existing situation. The problems here are twofold; the government must be able to formulate a policy consistent with society's 'national interest' and then must be able to implement that policy in its originally conceived structure. In the formulation and implementation stages of the government policy process, many bureaucratic and political pressures influence the development of policy. Politicians and government bureaucrats must be willing to subordinate personal ambition and self-interest objectives to the interests of society. This may, for instance, involve government bureaucrats implementing policies which reduce their powers and reduce the bureau's budget size. Similarly, politicians may be required to implement policies from which tangible benefits only accrue after that government's lifetime whilst costs are incurred in the short term.

In applying the economic theories of bureaucracies and politics to the development of UK oil policy since 1964 it is evident that the above conditions have not been fulfilled. Whether these conditions are likely to be fulfilled in the future depends largely on the structure of the policy process in the future. A crucial factor in the development of North Sea oil policy was its elevation to the forefront of public awareness by media attention at the time of the world oil price increases of 1973–4. The inherently unstable nature of the international oil market and the importance of oil revenues to the UK Treasury throughout the 1980s will tend to preserve North Sea oil policy as an important political issue. Groups and individuals active in the policy process bargain and negotiate for characteristics of policy. Pressures on the policy process derive from many sources which, although they may change in their relative bargaining strength over time, are likely to influence the oil policy process in the future.

Oil industry lobby organisations and oil companies exert influence on the oil policy process from outside government as do service and supply industry pressure groups and specialist groups such as the Institute for Fiscal Studies. Inside government the Foreign Office has an interest in North Sea oil policy to ensure that any government action is internationally acceptable (notably with regards to the European Community) and is not likely to provoke retaliation affecting British interests abroad (notably with regards to OPEC). The Treasury is primarily concerned with the macro-economic implications of oil policy and thus would have an interest in all aspects of oil policy, in particular with taxation and depletion policies. The Inland Revenue's direct concern with taxation policy is largely to do with the implementation of the tax system rather than the outcome of tax policy. The Department of Energy was created in January 1974 at least partly as a result of public attention surrounding the world oil 'crisis'.[7] Thus the existence of the Department of Energy can itself be seen as an aspect of oil policy whereby the government promoted energy to a higher status than had previously prevailed. The various groups and sections (see Figure 8.1) within the Department of Energy have specific duties and responsibilities and thus cannot be viewed as possessing a single set of objectives and priorities. The analysis of government oil policy since 1964 highlights the importance of political and bureaucratic pressures on the oil policy process. In providing an overview, it is important to note that groups active in the policy process desire characteristics of policy and derive utility from characteristics inherent in the whole policy process rather than solely from the policy output. For instance, some government bureaucrats (in the Treasury) may be primarily concerned with the output of oil taxation policy (the government take), whereas other government bureaucrats, say in the Inland Revenue, may be concerned with the policy input (the administrative complexity of the oil tax system). Both 'revenue' and 'complexity' are characteristics of oil taxation policy.

In order to analyse oil policy development using the intrinsic characteristics of policies it is necessary to distinguish between two broad stages in the policy process. Although the policy process is a continuous, dynamic process the two stages are employed as approximations for ease of presentation. The first stage refers to situations arising from the demand for 'new' policy. Demand for policy generally arises due to the political perception of need (see Chapter 3). In the very early stages of policy determination the politician is heavily reliant on the advice of government bureaucracies. The politician's reliance on the bureaucrat for information is especially important in technical and specialised areas where the decision-making politician is essentially a layman. The nature of the relationship between the sponsor (politician) and the bureaucrat (see Chapter 3) is such that the bureaucrat may have the ability to 'steer' the policy process. By selecting and withholding information the government bureaucrat may acquire characteristics of policy which are consistent with the bureaucrat's ambitions. Bargaining between the politician and bureaucrat takes place on the basis of characteristics of oil policies. At the initial policy formulation stage of the policy process the bureaucrat desires characteristics of oil policy which enable him to maximise

153

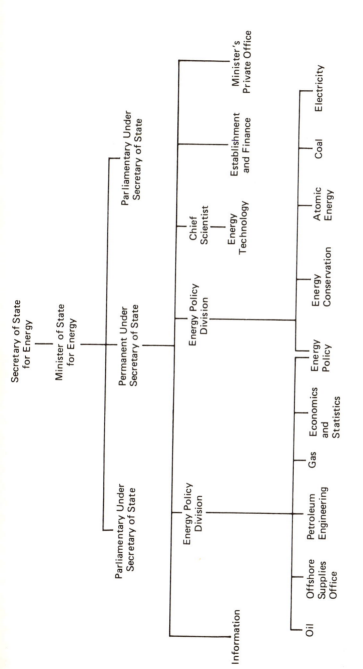

FIGURE 8.1 *Department of Energy, lines of responsibility*

budget size in the future (if that is the bureaucrat's objectives) i.e. the bureaucrat attempts to create the potential for budget maximisation. In stage two, when the policy framework is in existence, the bureaucrat may then attempt to maximise budget size (as Niskanen asserts; Chapter 3). The main difference between the two stages is that in stage one the budget size of a policy is a constraint on the bureaucrat's ambitions whereas in stage two the budget is itself a characteristic which the bureaucrat attempts to acquire more of.

In stage one the bureaucrat and the politician create the basic structure of the policy and it is within this structure that subsequent discussion and bargaining occurs. Initial policy discussions crucially affect the direction and nature of further policy developments. With respect to UK oil policy the decision in 1964 to adopt a discretionary licence allocation system had a considerable impact on oil policy decisions taken over a decade later. Thus the bureaucrat, in stage one of the policy process, attempts to introduce and emphasise characteristics of policy which create precedents and build a legal and institutional framework for future control and intervention. Functions and positions are created in order that over time flexibility will enable bureaucrats to take advantage of changing circumstances to acquire characteristics which directly maximise budget size or achieve other desirable bureaucratic objectives. UK oil depletion policy is characterised by the existence of the framework for depletion controls which have yet to be used extensively (see Chapter 6). Only since the legal potential for depletion controls came into existence have discussions concerning the implementation of controls occurred with lobby groups and independent analysts.

Thus in the short term (stage one of the policy process) the bureaucrat may attempt to alter the politician's perception of policy characteristics in order to maximise the potential for increases in budget size. In the longer term, the bureaucrat may attempt to alter the politician's perception of policy cost (the position of curve C in Figure 8.2) and the marginal value of policy (curve V) so as to move Line DF to the right and increase the size of Areas E and F. Throughout the bargaining process groups may thus emphasise or overstate key characteristics of policy for tactical purposes.

In stage two of the policy process a characteristic approach may, at least partly, explain the adoption (or non-adoption) of a new or differentiated policy. For a new policy to be adopted it must possess at least some of the characteristics desired by the major groups in the policy process. The non-adoption of the IFS oil tax proposals in

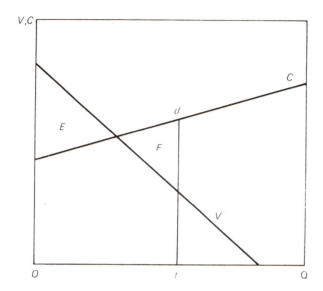

FIGURE 8.2 *The equilibrium output of a bureaucracy*

SOURCE W. Niskanen, *Bureaucracy: Servant or Master?* (London: IEA, 1973) p. 29.

December 1981 (see Chapter 7) could be partly explained in terms of their lack of characteristics (for example, administrative complexity) favoured by some government bureaucrats. A policy reducing administrative control and complexity could result in a decrease of budget size or workload for the bureaucrat thus, *ceteris paribus*, the bureaucrat is likely to oppose its introduction unless compensated by the inclusion in policy of another favourable characteristic.

In the consideration of more than two interest groups, characteristics and policies, the bargaining process becomes increasingly complex. At any point in time it may be possible to predict types of policy characteristics desired by various groups, and in the framework of the economic and political climate, forecast policy trends. At least it may be possible to predict that a policy is unlikely to be adopted because of certain characteristics of that policy being opposed by powerful groups in the policy process.

It is important to note that pressures are exerted on a policy during its passage through the policy process which may result in a final policy being an amalgam of a selection of policies. These pressures do not necessarily act as a restraint on extreme or unacceptable policies or activities of government. Bureaucratic and political influences on the policy process are not designed to achieve balance or moderation although this may occur. Pressures during the policy process are more likely to distort or bias policies rather than restrain policy output. Self-interest objectives are fulfilled through relative strength in bargaining and in influencing the policy process. Bureaucrats and politicians involved in developing policy are generally more aware of arguments put forward by producer pressure groups than those of consumer pressure groups (see Chapter 3). Thus because of their relatively powerful position in the policy process, producer pressure groups are more likely to achieve success (relative to consumer groups) in influencing the development of a policy. Moreover, the characteristics approach to policy development highlights the import-ance to various groups (particularly government bureaucrats) of the policy *per se* rather than the effects of the policy when implemented. Balance or restraint would be directed at policy output.

The intrinsic deficiencies of the government policy process cast doubts on the government's ability to intervene effectively in the oil industry. Irrespective of whether the government's policy intentions are desirable in terms of economic efficiency, the preceding analysis suggests government intervention is unlikely to fulfil the govern-ment's own objectives. It is accepted that there is a need for controls and state involvement in order to influence areas of the economy where the market system neglects certain social costs and benefits. Thus, the question arises: to what degree and in what form should government intervention in the offshore oil industry occur? Although this is essentially a normative question, a positive appraisal of North Sea oil policy may contribute to an assessment of the government's rôle in the offshore industry. Conflicts and inefficiencies may be understood and recognised as being inherent in the policy process thus assisting the development of forecasts and project evaluation.

A major virtue of the competitive market system is that it com-bines a tendency towards economic efficiency with a degree of indi-vidual freedom. At the same time, a drawback of the free market system is that over time it has been shown to lead to exploitation and abuse unless the government takes action to protect various sectors of society. Market rewards attract factors to produce at the lowest cost those goods which consumers most desire. Thus a desirable characteristic of North Sea oil policy is to attempt to create an environment in which, within certain constraints, distortions to the efficient allocation of scarce resources are reduced to a minimum.

Clarity of policy, with regard to the intentions and purpose of policy and the method of policy implementation, reduces uncertainty in the oil industry. Simplicity of policy structure enables policy to be monitored outside the government system, for instance, by independent academics or study groups, and in addition facilitates accountability within government.

The oligopolistic structure of the offshore oil industry in the UK, whilst encouraging the speedy exploitation of offshore resources, has given rise to various situations which the government has a clear function to rectify. Because of the strategic importance of North Sea oil both nationally and internationally the government must have access to information concerning the oil industry. The problem of gathering and interpreting information is a central issue in the theories of bureaucracies. A government bureaucracy with the function of gathering company and oilfield information for the purpose of ensuring 'good oilfield practice' may tend to increase its power and influence by selectively using its expertise to steer policy. Thus it is crucial for governments to state explicitly the limits of policies and the purposes of gathering information in order to constrain bureaucratic strategic behaviour. A degree of flexibility in policy is necessary in order to be able to respond to large-scale extraneous shocks to the oil market. Structural changes to policy in response to, say, increases in the world oil price, are indications of an unsatisfactory policy (an example is oil taxation policy, Chapter 7). However, unless the circumstances in which the government may invoke emergency powers are clearly identified in advance, their existence alone creates uncertainty.

With respect to the UK oil industry, there seems little evidence to support the view that the market mechanism cannot be expected to deal adequately with planning for the uncertainties of the future. The private sector has the expertise and experience to plan into the medium and long term. Past attempts by UK governments to influence the market have tended to distort market signals thus complicating and confusing oil company planning. Government intervention in the oil industry involves a movement away from employing the price mechanism in the allocation of scarce resources.

The UK system of representative democracy creates bureaucratic and political difficulties in the development of policy which are not obviously overcome within the existing system of government. North Sea oil in the 1970s became an important political issue and in the future politicians and political parties may be expected to use political and personal criteria in the continuing development and implementation of oil policy. Similarly, without radical constitutional reform the influence of the government bureaucrat in the policy process is likely to remain very powerful. Nevertheless, there are measures that may be introduced to control and constrain the activities of government bureaucrats.

Freedom of access to government information can make the bureaucrat more accountable to the public but the problem of measuring the efficiency of government output remains.[8] More importantly, freedom of information can increase competition in the government policy process. By giving independent organisations and individuals outside government access to information alternative policy proposals and their effects may be made clearer to politicians and to the public. Furthermore, a degree of duplication of policy within government similarly creates competition. Competition between government bureaucrats can lead to bureaucrats bidding for policies.[9] Bidding would be based on the characteristics of policy and competition would encourage the bureaucrat to include less of those characteristics that are directly designed to maximise the bureaucrat's utility function in policy proposals.[10] Moreover, by minimising the extent of government intervention in the offshore industry the potential and scope for political and bureaucratic self-interest to distort the market mechanism is reduced. In the area of the government policy process, two clear areas for future research are apparent as a result of the analysis in preceding chapters. First, a characteristic approach to the determination of policy may be developed in order to illustrate the bargaining process between interest groups within the government policy process. Secondly, there is large scope for examining the structure of government in the context of the economic theories of politics and bureaucracies.

In conclusion it is important to emphasise two points concerning government and North Sea oil policy. First, preceding analysis does not imply advocacy of the total withdrawal of government from the domestic oil industry. The government has many duties to perform in the North Sea as it does in all other sectors of the economy. These duties include safety control, pollution control, to guard against improper business practices and to protect offshore installations.[11]

Secondly, the government has a multiplicity of objectives to fulfil which include political, social and economic ambitions. On many occasions there may be a trade-off between various objectives and it is crucial to understand how, for example, economic objectives may be subjugated for political or ideological objectives. It is not suggested that economic efficiency in the strict sense should always take precedence over other objectives. However a politician or political party favours ideological or other commitments over, say, economic efficiency it is vital for governments in a democracy to make as explicit as they can what the economic costs and implications of policy are.

In highlighting the existence and sources of bureaucratic and political distortions in the policy process the shortcomings and deficiencies of government action may be recognised. UK North Sea oil policy since 1964 illustrates the practical limitations on the ability of governments to intervene effectively in the oil industry. Extending the rules and methodology of neo-classical economics to an examination of government oil policy contributes to an understanding of the nature of the imperfections in the government policy process. At least the potential for future distortions and the direction of those distortions may be estimated, thus assisting in forecasting and planning in the North Sea oil sector.

References

1 INTRODUCTION

1. C. Robinson, 'The Errors of North Sea Policy', *Lloyds Bank Review*, July 1981.
2. *The Petroleum Economist*, Vol. L, No. 3, March 1983, p. 116.

2 THE HISTORICAL BACKGROUND

1. C. Robinson, 'A Review of North Sea Policy', *Energie Wirtschaft*, April 1978.
2. Department of Energy, *Digest of UK Energy Statistics*, 1978.
3. Department of Energy, *Digest of UK Energy Statistics*, 1980.
4. C. Robinson, 'The Errors of North Sea Policy', *Lloyds Bank Review*, July 1981, No. 141, p. 14.
5. *North Sea Oil and Gas*, First Report from the Committee on Public Accounts, House of Commons, Session 1972–73 (London, HMSO, 1973) p. x.
6. O. Noreng, *The Oil Industry and Government Strategy in the North Sea* (London: Croom Helm, 1980) pp. 39–41.
7. First Report from the Select Committee on Nationalized Industries, Session 1974–75, HC 345 (London: HMSO, 1975) p. 193.
8. 'PAC First Report', p. 64.
9. Robinson, 'The Errors of North Sea Policy', p. 22.
10. Ibid, p. 24.
11. Ibid, p. 24.
12. C. Robinson and J. Morgan, *North Sea Oil in the Future, Economic Analysis and Government Policy* (London: Macmillan Press, 1978) chapter 1.
13. D. Meadows, *The Limits to Growth, A Report for The Club of Rome's Project on the Predicament of Mankind* (London: Earth Island, 1972).
14. *United Kingdom Offshore Oil and Gas Policy*, Cmnd. 5696 (London: HMSO, 1974).
15. Ibid, para. 4.
16. Ibid, para. 5.
17. Ibid, para. 5.
18. *1974 Labour Party Election Manifesto* (London: Labour Party, 1974).
19. *Hansard*, Vol. 891, Col. 536.

20. *Hansard*, Vol. 891, Col. 603.
21. HC 345, Q. 46.
22. *Hansard*, Vol. 896, Col. 1305–6.
23. *Hansard*, Vol. 896, Col. 1337–8.
24. *House of Commons, Written Answers*, 6 December 1974, Col. 650.
25. Robinson and Morgan, *North Sea Oil in the Future*, p. 131.
26. HC 345, Q. 46.
27. E. Krapels, *Controlling Oil: British Oil Policy and the British National Oil Corporation* (Washington: US Government Printing Office, 1977) p. 26.
28. *BNOC Report and Accounts for 1976* (London: BNOC, 1976).
29. *Seventh Report from the Select Committee on Nationalized Industries, Session 1977–78* (London: HMSO, 1978) Q. 194.
30. *BNOC Report*.
31. R. Bailey, 'BNOC's New Phase', *Nat-West Quarterly Review*, November 1979, p. 4.
32. S. Brittan in *The Financial Times*, 26 May 1977, and with B. Riley in the *Lloyds Bank Review*, April 1978.
33. C. Rowland, 'UK North Sea Tax Changes', *Noroil*, May 1982, p. 35.
34. C. Robinson, 'Oil Depletion Policy in the United Kingdom', *Three Banks Review*, No. 135, September 1982.
35. *Development of the Oil and Gas Resources of the United Kingdom, 1981* (London: HMSO, 1981) Appendix 9.
36. *The Financial Times*, 24 July 1980.
37. *Hansard*, 8 June 1982, Written Answers, p. 6.
38. *Oil and Gas (Enterprise) Act 1982* (London: HMSO, 1982).
39. *Offshore Engineer*, September 1982, p. 61.
40. *Oilman Newsletter*, 9 April 1983, p.3.
41. Robinson, 'The Errors of North Sea Policy', p. 21.

3 THE THEORETICAL FRAMEWORK

1. A. Downs, *An Economic Theory of Democracy* (New York: Harper & Row, 1957).
2. K. Arrow, *Social Choice and Individual Values* (New York: John Wiley, 1963).
3. A. Breton, *The Economic Theory of Representative Government* (Chicago: Aldine Publishing Co., 1974) p. 3.
4. Downs, *Economic Theory of Democracy*, p. 247.
5. G. Tullock, 'Some Problems of Majority Voting', *Journal of Political Economy*, December 1959.
6. G. Tullock, *Towards a Mathematics of Politics* (Ann Arbor: University of Michigan Press, 1967) p. 50.
7. A. Downs, 'In Defense of Majority Voting', *Journal of Political Economy*, April 1961.
8. G. Tullock and J. Buchanan, *The Calculus of Consent* (Ann Arbor: University of Michigan Press, 1962).

9. G. Tullock, *The Vote Motive* (London: Institute of Economic Affairs, 1976).
10. Ibid, p. 51.
11. Tullock and Buchanan, *Calculus of Consent*.
12. Tullock, *Towards a Mathematics of Politics*, p. 50.
13. W. Riker, *The Theory of Political Coalitions* (New Haven and London: Yale University Press, 1962).
14. Tullock, *Towards a Mathematics of Politics*, pp. 52–3.
15. Tullock, *The Vote Motive*, p. 23.
16. Breton, *Economic Theory of Representative Government*, chapter 3.
17. Ibid, p. 5.
18. Ibid, p. 7.
19. Ibid, chapter 4.
20. Ibid, chapter 7.
21. Ibid, p. 143.
22. Tullock, *The Vote Motive*, p. 29.
23. A. Downs, *Inside Bureaucracy* (Boston, Mass.: Little, Brown, 1967).
24. W. Niskanen, *Bureaucracy and Representative Government* (Chicago: Aldine-Atherton, 1971); *Bureaucracy, Servant or Master?* (London: Institute of Economic Affairs, 1973); 'Bureaucrats and Politicians', *Journal of Law and Economics*, December 1975; 'The Peculiar Economics of Bureaucracy', *American Economic Review*, May 1968, and also in *The Economics of Property Rights* by E. Furubotn and B. Pejovich (Cambridge Mass.: Ballinger Publishing Co., 1974).
25. Niskanen, *Bureaucracy and Representative Government*, chapter 5.
26. Breton, *Economic Theory of Representative Government*, p. 163.
27. Niskanen, *Bureaucracy and Representative Government*, p. 45.
28. Ibid, p. 49.
29. A. Breton and R. Wintrobe, 'The Equilibrium Size of a Budget Maximizing Bureau: A Note on Niskanen's Theory of Bureaucracy', *Journal of Political Economy*, February 1975.
30. Ibid, p. 198.
31. D. Mueller, *Public Choice* (Cambridge: Cambridge University Press, 1980) p. 158.
32. Breton and Wintrobe, 'The Equilibrium Size of a Budget Maximizing Bureau', p. 205.
33. Breton, *Economic Theory of Representative Government*, chapter 10.
34. A. Breton and R. Wintrobe, *The Logic of Bureaucratic Conduct* (Cambridge: Cambridge University Press, 1982).
35. J. Buchanan, 'Politics, Policy, and the Pigovian Margins', *Economica*, 1962.
36. *Hansard*, Vol. 896, Col. 1305–6.

4 LICENSING POLICY

1. K. Dam, *Oil Resources, Who Gets What How?* (London: University of Chicago Press, 1976) p. 34.

2. Ibid, p. 36.
3. Ibid, p. 43.
4. J. B. Ramsey, *Bidding and Oil Leases* (Greenwich, Connecticut: Jai Press, 1980) pp. 137–8.
5. *Oilman*, October 1982, p. 9.
6. C. Robinson and J. Morgan, *North Sea Oil in the Future, Economic Analysis and Government Policy* (London: Macmillan, 1978) p. 194.
7. Dam, *Oil Resources, Who Gets What How?* p. 10.
8. Robinson and Morgan, *North Sea Oil in the Future*, p. 195.
9. First Report from the Select Committee on Nationalized Industries, Session 1974–5, *Nationalized Industries and the Exploitation of North Sea Oil and Gas*, HC 345 (London: HMSO, 1975) p. 193.
10. First Report from the Committee on Public Accounts, Session 1972–3, *North Sea Oil and Gas* (London: HMSO, 1973) p. 26.
11. Ibid, p. 74.
12. Dam, *Oil Resources. Who Gets What How?* p. 26.
13. HC 345, p. 193.
14. HC 345, p. 193.
15. Dam, p. 29.
16. PAC First Report, para. 34.
17. PAC First Report, para. 43.
18. PAC First Report, Q. 113–14.
19. PAC First Report, Q. 16.
20. P. Odell, *British Oil Policy: A Radical Alternative* (London: Kogan Page, 1979).
21. PAC First Report, pp. 152–3.
22. Robinson and Morgan, *North Sea Oil in the Future*, p. 194.
23. A. Hamilton, *North Sea Impact, Offshore Oil and the British Economy* (London: International Institute for Economic Research, 1978) p. 108.
24. Department of Energy, *United Kingdom Offshore Oil and Gas Policy*, Cmnd 5696 (London: HMSO, 1974).
25. Ibid, p. 5.
26. Ibid, p. 5.
27. A. Hamilton, *North Sea Impact*, p. 43.
28. Department of Energy, *Development of the Oil and Gas Resources of the UK, 1977* (London: HMSO, 1977).
29. *The Economist*, 12 February 1977, p. 105.
30. *Hansard*, Vol. 891, Col. 492.
31. R. Hallyer and R. Pleasance, *UK Taxation of Offshore Oil and Gas* (London: Butterworths, 1977) p. 10.
32. Ibid, p. 12.
33. *Oil and Gas Journal*, 15 May 1978, p. 50.
34. C. Johnson, *North Sea Energy Wealth 1965–1985* (London: Financial Times, 1979) p. 9.
35. C. Robinson, 'The Errors of North Sea Policy', *Lloyds Bank Review*, July 1981.
36. *Daily Telegraph*, 11 December 1978, p. 7.
37. *The Economist*, 25 November 1978, p. 11.

38. Department of Energy, *Development of the Oil and Gas Resources of the UK, 1980* (London: HMSO, 1980).
39. Department of Energy, *Development of the Oil and Gas Resources of the UK, 1981* (London: HMSO, 1981) p. 5.
40. 'North Sea Bids Offer £175m Bonus', *Financial Times*, 12 August 1980.
41. *Noroil*, October 1982, p. 25.
42. *Oilman*, 2 October 1982.
43. Ramsey, *Bidding and Oil Leases*, chapter 7.
44. Dam, *Oil Resources. Who Gets What How*, chapter 4.
45. Ibid.
46. Ibid, p. 149.
47. Ramsey, *Bidding and Oil Leases*, p. 161.
48. A. Clunies Ross, 'North Sea Oil and Gas Taxation', *Three Banks Review*, June 1982, No. 134.

5 PARTICIPATION

1. *North Sea Oil and Gas*, First Report from the Committee of Public Accounts, House of Commons, Session 1972–3, p. 26.
2. C. Robinson and J. Morgan, *North Sea Oil in the Future, Economic Analysis and Government Policy* (London: Macmillan, 1978) chapter 4.
3. L. Grayson, *National Oil Companies* (New York: John Wiley, 1981) p. 225.
4. R. Normann and E. Rhenman, *Formulation of Goals and Measurement of Effectiveness in the Public Administration* (Stockholm: SIAR, 1975) p. 51.
5. *United Kingdom Offshore Oil and Gas Policy*, Cmnd 5696 (London: HMSO, 1974).
6. Ibid, para. 4.
7. A. Downs, *An Economic Theory of Democracy* (New York: Harper & Row, 1957).
8. A. Breton, *The Economic Theory of Representative Government* (Chicago: Aldine Publishing Co., 1974).
9. Cmnd 5696.
10. *The Economist*, 1 November 1975, p. 95.
11. G. Tullock, *Towards a Mathematics of Politics* (Ann Arbor: University of Michigan Press, 1967); W. Riker, *The Theory of Political Coalitions* (New Haven and London: Yale University Press, 1962).
12. A. Downs, *Economic Theory of Democracy*; and G. Tullock, *The Vote Motive* (London: IEA, 1973) pp. 24–5.
13. *Hansard*, Vol. 891, Col. 603.
14. First Report from the Select Committee on Nationalized Industries, Session 1974–5, *Nationalized Industries and the Exploitation of North Sea Oil and Gas*, HC 345 (London: HMSO, 1975) Q. 633.
15. *Hansard*, Vol. 896, Col. 1304–6.
16. *Hansard*, Vol. 896, Col. 1337–8.
17. O. Noreng, *The Oil Industry and Government Strategy in the North Sea*

(London: Croom Helm, 1980) p. 134.
18. *Hansard*, Written Answers, 6 December 1974, Col. 650.
19. W. Niskanen, *Bureaucracy and Representative Government* (Chicago: Aldine-Atherton, 1971).
20. W. Niskanen, *Bureaucracy; Servant or Master?* (Wolverhampton: IEA, 1973) p. 23.
21. Grayson, *National Oil Companies*, chapter 1.
22. *Hansard*, Vol. 891, Col. 490.
23. *Hansard*, Vol. 891, Col. 499.
24. Downs, *Economic Theory of Democracy*, p. 255.
25. Breton, *Economic Theory of Representative Government*, p. 81.
26. *Petroleum and Submarine Pipelines Act, 1975* (London: HMSO, 1975) p. 3.
27. *Financial Times*, 30 June 1981, p. 18.
28. *Petroleum and Submarine Pipelines Act, 1975*, p. 6.
29. *The Economist*, 12 July 1975, p. 72.
30. Robinson and Morgan, *North Sea Oil in the Future*, pp. 30–1.
31. *Hansard*, Written Answers, 30 April 1975.
32. *Hansard*, Vol. 896, Col. 1427–8.
33. Niskanen, *Bureaucracy and Representative Government*.
34. E. Krapels, *Controlling Oil: British Oil Policy and the British National Oil Corporation* (Washington: US Government Printing Office, 1977) p. 26.
35. *BNOC Report and Accounts, 1976* (London: BNOC, 1977).
36. *Seventh Report from the Select Committee on Nationalized Industries*, Session 1977–8, HC 583 (London: HMSO, 1978) Q. 223.
37. K. Dam, *Oil Resources, Who Gets What How?* (Chicago: University of Chicago Press, 1976) chapter 12.
38. Grayson, *National Oil Companies*, chapter 6.
39. G. Corti and F. Frazer, *The Nation's Oil, A Story of Control* (London: Graham & Trotman, 1983) p. 140.
40. Ibid, chapter 7.
41. HC 583, Q. 187.
42. HC 583, Q. 189.
43. *Oil and Gas Journal*, 7 March 1977, p. 83.
44. HC 583, Q. 201.
45. Corti and Frazer, *The Nation's Oil*, p. 165.
46. *BNOC Report and Accounts, 1977* (London: BNOC, 1978).
47. Grayson, *National Oil Companies*, p. 188.
48. A. Breton and R. Wintrobe, 'The Equilibrium Size of a Budget-Maximizing Bureau: A Note on Niskanen's Theory of Bureaucracy', *Journal of Political Economy*, February 1975.
49. Breton, *Economic Theory of Representative Government*.
50. Grayson, *National Oil Companies*, chapter 12.
51. Niskanen, *Bureaucracy and Representative Government*, chapter 2.
52. Grayson, *National Oil Companies*, p. 254.
53. O. E. Williamson, 'A Model of Rational Managerial Behaviour', in R. Cyert and J. March, *A Behavioural Theory of the Firm* (Englewood Cliffs: Prentice-Hall, 1963).

166 *References*

54. Ibid.
55. *Financial Times*, 30 June 1981, p. 18.
56. *The Economist*, 25 July 1981, p. 29.
57. By Labour Ministers and by Lord Kearton, e.g. *The Times*, 16 February 1978, p. 15.
58. *BNOC Report and Accounts for 1979* (London: BNOC, 1980).
59. C. Robinson, 'The Errors of North Sea Policy', *Lloyds Bank Review*, July 1981, p. 18.
60. *The Economist*, 30 June 1979, p. 77.
61. *Oil and Gas Journal*, 30 July 1979, p. 118.
62. *Daily Telegraph*, 27 July 1979, p. 19.
63. *The Economist*, 11 August 1979.
64. *BNOC Report and Accounts for 1979* (London: BNOC, 1980).
65. *The Times*, 28 October 1982, p. 11.
66. *The Times*, 28 October 1982, p. 10.
67. *Oil and Gas (Enterprise) Act 1982* (London: HMSO, 1982).
68. *BNOC Report and Accounts for 1982* (London: BNOC, 1983).
69. P. Stevens, *The International Oil Market*, unpublished mimeographed paper, University of Surrey, 1983.
70. *The Sunday Times*, 14 November 1982, p. 61.
71. *Financial Times*, 1 November 1982, p. 14.
72. *The Times*, 20 November 1982, p. 1.
73. *Noroil*, December 1982, p. 39.
74. *Financial Times*, 20 October 1981, p. 1.
75. *Financial Times*, 26 November 1982, p. 23.

6 DEPLETION POLICY

1. C. Robinson and J. Morgan, *North Sea Oil in the Future, Economic Analysis and Government Policy* (London: Macmillan, 1978) chapter 2.
2. Ibid.
3. *North Sea Oil and Gas*, First Report from the Committee of Public Accounts, House of Commons, Session 1972–3, (London: HMSO, 1973) para. 96.
4. *1964 Continental Shelf Act* (London: HMSO, 1964); *Petroleum (Production) Regulations*, 1964, 1966, 1971.
5. D. Meadows, *The Limits to Growth: A Report for the Club of Rome's Project on the Predicament of Mankind* (London: Earth Island 1972).
6. C. Robinson, 'Oil Depletion Policy in the United Kingdom', *Three Banks Review*, No. 135, September 1982, p. 9.
7. Ibid, p. 10.
8. *Energy Commission Paper #17*, UKOOA, Table BII, p. A34.
9. Department of Energy, *Development of the Oil and Gas Resources of the UK, 1983* (London: HMSO, 1983) Appendix 6.
10. Robinson and Morgan, *North Sea Oil in the Future*, chapter 2.
11. PAC First Report, Q. 304.
12. *OPEC Annual Statistical Bulletin 1977*, pp. 129–30.

13. *UK Offshore Oil and Gas Policy*, Cmnd 5690 (London: HMSO, 1974).
14. Ibid, p. 5.
15. *Financial Times*, 28 May 1974.
16. Ibid, 12 September 1974.
17. A. Downs, *An Economic Theory of Democracy* (New York: Harper & Row, 1957).
18. A. Breton, *The Economic Theory of Representative Government* (Chicago: Aldine Publishing Co., 1974).
19. Downs, *Economic Theory of Democracy*.
20. G. Tullock, *The Vote Motive* (London: Institute of Economic Affairs, 1976).
21. Ibid, p. 24.
22. Downs, *Economic Theory of Democracy*, p. 247.
23. House of Commons Select Committee on Energy, Session 1980–81, *Minutes of Evidence: North Sea Oil Depletion Policy*, HC 321 C. Robinson, p. 240 (London: HMSO, 1981).
24. Ibid, HC 321(i), Q. 129; and also HC 321(ii), Q. 149.
25. *Petroleum and Submarine Pipelines Act, 1975* (London: HMSO, 1975).
26. *Hansard*, Written Answers, 6 December 1974, Cols 648–50.
27. Robinson and Morgan, *North Sea Oil in the Future*, pp. 30–1.
28. Ibid, p. 33.
29. Breton, *Economic Theory of Representative Government*, p. 163.
30. HC 321(i), p. 41.
31. *Hansard*, Written Answers, 6 December 1974, Vol. 882, Col. 650.
32. *Oil and Gas Journal*, 30 July 1979, p. 118.
33. Department of Energy, *Development of Oil and Gas Resources of the United Kingdom 1981* (London: HMSO, 1981) Appendix 9.
34. Ibid, para. 2.
35. Ibid, para. 2.
36. Ibid, para. 3.
37. Ibid, para. 7.
38. Ibid, para. 3.
39. HC 321(iii), Q. 317.
40. HC 321(iii), Q. 315.
41. Select Committee Report.
42. Ibid, P Odell.
43. HC 321(iii), Q. 307.
44. Ibid, Q. 283.
45. Ibid, Q. 307.
46. Select Committee Report.
47. HC 321(iii), p. 138.
48. HC 321(i), Q. 124.
49. HC 321(iii), p. 142.
50. *Financial Times*, 23 April 1982.
51. Select Committee Report, para. 58.
52. Ibid, para. 74.
53. Ibid, para. 65.
54. Robinson and Morgan, *North Sea Oil in the Future*, chapter 2.
55. HC 321(ii), Q. 254.

56. Robinson, 'Oil Depletion Policy in the UK', p.14.
57. Select Committee Report, para. 92.
58. *Guardian*, 9 June 1982, p. 17.
59. C. Robinson, evidence to the Select Committee, HC 321, Q. 494.

7 TAXATION POLICY

1. *North Sea Oil and Gas*, First Report from the Committee of Public Accounts, House of Commons, Session 1972–3 (London: HMSO, 1973).
2. Ibid.
3. Ibid, para. 96.
4. Ibid, para. 73.
5. Ibid, para. 66.
6. Ibid, para. 57.
7. Ibid, para. 58.
8. Ibid, para. 59.
9. Ibid, para. 60.
10. Ibid, para. 66.
11. Ibid, para. 98.
12. Ibid, para. 62.
13. Ibid, para. 62.
14. Ibid, Annex 12, p. 56.
15. Ibid, Q. 190–210.
16. Ibid, para. 70.
17. R. Esam, 'A View from the Industry', Institute for Fiscal Studies Conference: *The Taxation of North Sea Oil*, February 1976, p. 53.
18. Ibid.
19. Ibid, p. 54.
20. J. Davis, *High Cost Oil and Gas* (London: Croom Helm, 1981) p. 103.
21. G. Brennan and J. Buchanan, 'Towards a Tax Constitution for Leviathan', *Journal of Public Economics*, 1977, pp. 255–73.
22. PAC First Report.
23. Ibid, para. 66.
24. *UK Offshore Oil and Gas Policy*, Cmnd 5696 (London: HMSO, 1974).
25. Ibid, p. 4.
26. Ibid, p. 8.
27. Ibid, p. 4.
28. Esam, 'A View from the Industry', p. 56.
29. C. Robinson and J. Morgan, *North Sea Oil in the Future, Economic Analysis and Government Policy* (London: Macmillan, 1978) p. 85.
30. Cmnd 5696.
31. W. Niskanen, *Bureaucracy and Representative Government* (Chicago: Aldine-Atherton, 1971).
32. Robinson and Morgan, *North Sea Oil in the Future*, chapter 5.
33. Ibid, p. 86.
34. Ibid, p. 86.
35. K. Dam, *Oil Resources, Who Gets What How?* (London: University of

Chicago Press, 1977) p. 125.
36. E. Dell, *Hansard*, Vol. 886–7, p. 290, 25 February 1975.
37. Inland Revenue Press Release, 16 January 1975, p. 1.
38. Dell, *Hansard*, 25 February 1975.
39. Dam, *Oil Resources, Who Gets What How?*, p. 127.
40. Dam, *Oil Resources, Who Gets What How?*, p. 126.
41. *UKOOA Report to the UK Government on the Proposed Increase to Petroleum Revenue Tax*, United Kingdom Offshore Operators Association, 15 January 1979, p. 7.
42. R. Hallyer and R. Pleasance, *UK Taxation of Offshore Oil and Gas* (London: Butterworth & Co., 1977) chapter 21.
43. Treasury Press Release, 25 February 1975.
44. A. Breton, *The Economic Theory of Representative Government* (Chicago: Aldine, 1974).
45. E. Dell, UKOOA Report, p. 10.
46. P. Jenkin, UKOOA Report, p. 10.
47. Treasury Press Release, 25 February 1975.
48. UKOOA Report, p. 11.
49. Ibid, p. 12.
50. Robinson and Morgan, *North Sea Oil in the Future*, p. 108.
51. A. Kemp and D. Cohen, *The New System of Petroleum Revenue Tax*, University of Aberdeen, December 1979, p. 1.
52. J. Barnett, Treasury Press Release, 2 August 1978.
53. J. Morgan, 'The Promise and Problems of PRT', Institute for Fiscal Studies Conference, February 1976.
54. H. Motamen and R. Strange, *The Structure of the UK Oil Taxation System*, Imperial College Discussion Papers, No. 82/2.
55. C. Johnson, *North Sea Energy Wealth 1965–1985* (London: Financial Times, 1979), p. 82.
56. *OPEC Bulletin*, May 1980, p. 22. Table I.
57. *Hansard*, Col. 1464–5, March 1980.
58. *Hansard*, Col. 1464.
59. Inland Revenue Press Release, 24 November 1980.
60. A. Kemp and D. Rose, 'Budget Changes in Oil Taxation', *Petroleum Economist*, April 1981.
61. Ibid, p. 147.
62. Ibid, p. 147.
63. *UK Offshore Tax*, UKOOA Report, November 1981.
64. Kemp and Rose, *New System of PRT*.
65. *Submission to the Chancellor of the Exchequer on the 1981 Tax Changes to the Oil Tax Regime*, UKOOA, October 1981.
66. *Oil Taxation in the United Kingdom: Submission to the Chancellor of the Exchequer*, BRINDEX, September 1981.
67. *The Taxation of North Sea Oil*, Report of a Committee set up by the Institute, Institute of Fiscal Studies, December 1981.
68. C. Rowland, 'UK North Sea Tax Changes', *Noroil*, May 1982, p. 35.
69. Ibid.
70. *Taxation of North Sea Oil*.
71. Ibid, p. 4.

72. A. Kemp and D. Rose, 'How to Keep Oil Flowing When the Price Moves Down', *Financial Times Energy Economist*, April 1983, Issue 18.
73. 'Budget Changes in Oil Taxation', *Noroil*, April 1983.
74. Ibid.
75. *Offshore Engineer*, June 1982, p. 20.
76. C. Rowland, 'UK North Sea Tax Changes', *Noroil*, May 1982, p. 37.
77. A. Kemp and D. Rose, 'North Sea Economics Revised', *Petroleum Economist*, April 1982.
78. Kemp and Rose, 'How to Keep Oil Flowing'.

8 AN OVERVIEW

1. K. Dam, *Oil Resources, Who Gets What How?* (London: University of Chicago Press, 1976) p. 25.
2. *North Sea Oil and Gas*, First Report from the Committee on Public Accounts, House of Commons, Session 1972–73 (London: HMSO, 1973).
3. *The New Hope for Britain, Labour's Manifesto 1983* (London: Labour Party, 1983), p. 15.
4. C. Robinson and J. Morgan, *North Sea Oil in the Future, Economic Analysis and Government Policy* (London: Macmillan, 1978) p. 43.
5. *Petroleum and Submarine Pipelines Act, 1975* (London: HMSO, 1975).
6. C. Robinson, 'The Errors of North Sea Policy', *Lloyds Bank Review*, July 1981.
7. L. Pearson, *The Organization of the Energy Industry* (London: Macmillan, 1981) p. 24.
8. R. McKenzie and G. Tullock, *The New World of Economics* (Homewood: Irwin, 1975) p. 207.
9. A. Breton, 'Economics of Representative Democracy', in J. Buchanan *The Economics of Politics*, IEA Readings, 18 (Lancing: Institute of Economic Affairs, 1978) p. 61.
10. A. Breton and R. Wintrobe, *The Logic of Bureaucratic Conduct* (London: Cambridge University Press, 1982).
11. Robinson, 'Errors of North Sea Policy'.

Index

171

172   *Index*

Conservatives
  in Government, 9, 14, 18, 47, 50,
    63–8, 79, 92–100,113, 120, 123
  in Opposition, 11, 18, 45, 75, 85,
    89, 105, 113, 131, 134
Continental Shelf Act 1964, 1, 5,
    11, 101
Corporation Tax, 124–5, 130, 139

Dam, K., 50
Dell, E., 129, 132–3
Deminex, 83
Department of Energy, 15, 140,
    152–3
  and depletion, 106, 108–12, 115–
    22
  ·and licensing, 49, 58, 62–3
  and participation, 77, 79, 84–8, 97
Department of Trade and Industry,
    104, 126
  and depletion, 12, 20, 77–8, 92,
    101–22
  and BNOC, 120
  and taxation, 119
  controls, 105, 110, 117, 119–20
  discretionary power, 109–14
  government attitude, 102, 120
  information, 116–7
  optimal rate, 103–117
  repletion, 115–120
  royalty banking, 120
discount rates, 9, 108
Downs, A., 1, 24–35, 44, 68, 72,
    75–6, 105–6, 110, 150
Dunlin Field, 15, 77

economic rent, 8, 47, 57, 69, 143
Ekofisk Field, 6, 55
Energy Act 1976, 110–11
'energy gap', 9
energy policy, 14, 69, 148
Energy Resources Conservation
    Board of Alberta, 12
Energy Select Committee, 115–16,
    119, 120, 123
enhanced oil recovery, 139
ENI, 69
Erroll, F., 143
Esso, 56, 62, 86, 117–18, 139

European Community, 151
externalities, 103

Finance Act 1979, 134
Finance Act 1981, 135
Foreign Office, 51, 151
Forties Field, 4, 7, 104
freedom of information, 158

Gas Council, 8, 49, 50, 52–4
government, 142, 146–59
  accountability, 131, 146–7, 157
  competition, 158
  control, 12–13, 69, 74, 139–40
  decision-making, 24, 93–100, 137
  departments, 146, 148
  intervention, 44, 68–101, 104,
    108, 143, 150
  tax revenues, 13, 19, 120, 127,
    134–6, 151–2
government, representative, 31–3,
    40, 66, 106, 150
Grayson, L., 69
Gulf, 52, 77
Gulf War, 134

Hamilton Brothers, 86
Hewett Field, 4
Howe, G., 135
Howell, D., 19, 20, 94–8, 113–15
Hutton Field, 77

Kearton, Lord, 15, 58, 72, 77, 81–
    92
Kerr McGee, 86

incremental investment, 139
indefatigable field, 4
Inland Revenue, 51, 126–7, 140,
    145, 151–2
Institute for Fiscal Studies, 136–7,
    151, 154
Iran, 93, 133–4
Irish Sea, 53–4

Jenkin, P., 76, 132

Labour, 6, 53–4, 65, 79, 89, 99, 11(
  Election Manifesto 1974, 11